Chromosomes,

Giant Molecules,

and Evolution

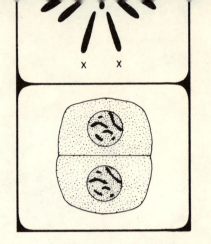

CHROMOSOMES,

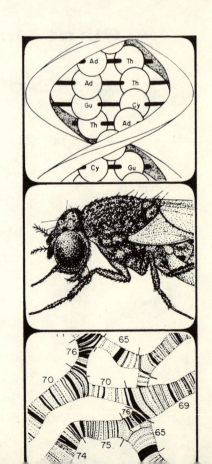

GIANT MOLECULES,

and EVOLUTION

by Bruce Wallace

CORNELL UNIVERSITY

ILLUSTRATED BY FRANCES ANN McKITTRICK

W·W·NORTON & COMPANY·INC·  NEW YORK

To
Natasha and Dodik

Contents

Not from that day to this has a single fact been discovered to shake the conclusions of Darwin that all living beings have evolved from earlier simpler forms; rather the mass of cumulative evidence has grown mountain high, so that no intelligent man can possibly deny to-day the fact *of organic evolution.*

J. H. RANDALL, JR.
The Making of the Modern Mind

Preface

The traditional approach to an understanding of evolution, the approach used by Darwin as well as by most present-day biology instructors, is based on the comparison of skulls, teeth, limbs, flower parts, and other anatomical features; on comparative embryology; on the fossil record; and on geographic variation. These, too, are the only sources of evidence with which most adults are aware.

Genetics has made extremely important contributions to evolutionary theory over the past three decades. Because of the brilliance of the pioneer evolutionary geneticists—R. A. Fisher, J. B. S. Haldane, and Sewall Wright—these contributions exist largely in mathematical terms; the atomistic nature of Mendelian inheritance, as these men appreciated so well, lends itself to precise quantitative descriptions that can be readily subjected to mathematical manipulation. Although students with mathematical ability can easily grasp these genetic arguments, older persons for whom algebra merely recalls a vague, bygone misery are generally unable or unwilling to master them.

Genetics contributes to the study of evolution, though, beyond merely supplying data that can be subjected to mathematical analysis. Certain genetic evidence leads unavoidably to the conclusion that evolution has occurred; the logic involved in interpreting this additional evidence is as rigorous as mathematical reasoning, but formal mathematical symbols are unnecessary. It is this type of evidence, evidence obtained both from gross chromosomal studies and from refined chemical analyses, that is summarized in the pages that follow. The presentation has been made as nontechnical as possible so that nonbiologists—students and their elders alike—can appreciate the observations and their interpretation. Thus, the terms *mitosis* and *meiosis,* for example, have not been used, and the details of *meiosis* have been stripped to the minimum that is essential for the purpose at hand. Accuracy, however, has not been unduly sacrificed; the genetics given here is sound.

An exceptionally large number of figures have been prepared to accompany the relatively brief text. Words are inadequate to describe some of the physical rearrangements of chromosomal parts, let alone the pairing configurations to which these can give rise. References to a single figure may occur here and there throughout the text as well as in the legends of other figures; consequently, all figures and their legends have been gathered together for convenience in a single section at the back of the book.

A number of persons have read and criticized the text during its preparation. It is a pleasure to thank my colleagues Professors Th. Dobzhansky, William Keeton, Charles Sibley, Adrian Srb, and Harry Stinson, Drs. Lee Ehrman, Ross MacIntyre, and Henry Shaffer, and Mr. John Puhalla for their comments and criticisms. Special appreciation must be expressed for the efforts of Professor Gerd Korman who played the taxing role of the intelligent layman in voicing his criticisms. Sins of either omission or commission, of course, are the author's.

BRUCE WALLACE

Haren (Gr.), the Netherlands
November 15, 1965

Chromosomes,

Giant Molecules,

and Evolution

1

Evolution Today: Divided Opinions

In a recent issue of *The American Scientist* a reviewer expressed his pleasure regarding a textbook of evolution in this manner: "It is free from the belabored proofs that evolution has happened, which take up so much of other works in this field." These are the 1960s. Some years ago centennial celebrations honoring the appearance of Darwin's *On the Origin of Species* were held throughout the world. Great men—the leading biologists of this generation—met to pay homage to the man whose mind had finally grasped the significance of regional variation among living things and of racial differences, the man who saw that all groups of organisms—diverse and complex as differences between them may seem —could have arisen through evolution from earlier forms. The evolution of life and its diversification into numerous species were, in Darwin's view, ascribable to natural selection, to the differential reproductive success of nonidentical individuals. There was, in this view, no need to account for each species by an act of special, and Divine, creation.

More than one hundred years after the formulation of a theory of evolution based on natural selection, our reviewer found a textbook willing to discuss the *process* of evolution without a long preamble in which the author justifies his belief that evolution has actually occurred. By this time, too, the biological world has, by seminars, symposia, and panel discussions, exhibited its awareness of the validity of Darwin's deductions. It would appear that biology has at last reached the stage in studies on evolution where real intellectual progress can be made, where each detailed and specialized advance need not be bogged down with a recitation of ABCs, where a variety of problems can be examined and questions posed—with no apology or delay—in the context of historical evolution.

At the very moment when the study of evolution appears finally to have broken out of its age-old restraints, however, confusion and dissension appear, too, more imminent than at any recent time. One hears that an effort has been made to restrict the teaching of evolution in the public schools of California to its presentation as a mere possibility, not as a proven fact. Although this is the position held by many persons of many faiths who do not accept evolution, it was spelled out most precisely by Pope Pius XII in 1953 when he stated:

In recent publications on genetics, one reads that nothing explains better the connection between living things than the picture of a common genealogical tree. But at the same time, one must note that this is nothing but a picture, a hypothesis, and not a demonstrated fact. One must add that if most scientists present the doctrine of evolutionary descent as a fact, this constitutes a hasty judgement. Other hypotheses can be formulated just as well. We can add furthermore that some well-known scientists do in fact present other suggestions without contesting that life has evolved, and that certain discoveries can be interpreted as forerunners of the human body. But, we add, these scientists have very clearly stressed that to their knowledge one does not yet know the precise meaning of the expressions "evolution," "descent," and "transition"; that one knows of no natural process by which one being could produce another of a different nature; that the process by which one species gives birth to another remains entirely impenetrable despite numerous intermediate stages; that no one has as yet succeeded experimentally in getting one species from another; and, finally, that one could never know definitely at which moment during evolution the human-like being suddenly passed the threshold of humanity.

Reports more drastic than those from California come from Arizona. Here it appears that an influential Baptist minister is waging a campaign to prohibit the teaching of evolution in the public schools of that state. Teachers who defied the prohibition would be fined and would lose both their certificates and their jobs. The Arizona campaigners can point with glee to the antievolution law that is still valid in Tennessee forty years after the notorious Scopes trial. Although still on Tennessee's books, this prohibition has not been enforced vigorously in recent years. At the time of this writing, South Carolina for the first time in more than one hundred years had permitted the use of the terms "Darwin" and "evolution" in textbooks adopted by state schools below the college level.

Finally, a Creation Research Society has been formed that will eventually, according to its brochure, publish a journal as well as textbooks for high-school and college use. Members of the society subscribe to the following beliefs: (1) that biblical accounts as given in the original autographs are true in all details and (2) that man, as well as other species of living things, was created in six days as described in the first book of the Bible.

The public, then, will soon be confronted with a bewildering array of reports. In some, evolution will be denied— outright and vehemently. In others, evolution will be an

3

implicit assumption—not actually spelled out but necessary for a real understanding of the text itself. Explanatory statements will be issued by those, among others, whose formal education was completed three or four decades ago; among these will be those more than casually interested in establishing a hierarchy of superiority among races of human beings. Once more we will be subjected to a debate on the perfection of the human eye, the meaning of embryonic similarities among different classes of animals, the interpretation of the fossil record, and the hidden significance of vestigial organs. In the hands of experts (including Darwin) taxonomic, anatomical, and embryological evidence constitutes the basis of a powerful argument; in the hands of the inexperienced, deprived of the logic that weaves a convincing case for evolution, the same facts can constitute a parody. Clever opponents of evolution—whatever their motivation—can wreak havoc with the parody.

In the chapters that follow I intend to present evidence for evolution that arises from modern genetics. As a rule genetic evidence is much easier to grasp than evidence based on anatomy, embryology, or taxonomy; the arguments accompanying the presentation of genetic evidence are generally less sophisticated than the arguments mustered by Darwin and the evolutionists who followed him. These arguments are very similar indeed to those used in courts of law where they often determine the life or death of the accused.

Before discussing either the arguments or the evidence, we must concede that the extreme antievolution viewpoint cannot be refuted. He who is unalterably convinced at the outset that the different species of plants and animals did indeed arise through separate acts of special creation by an Intelligent Being will find nothing in this book compelling him to believe otherwise. The reason is simple. This Intelligent Being can be imagined to have carried out His creative activities in a manner that duplicates what would otherwise appear to be an evolutionary scheme. Worse, one can imagine that He created false evidence to confuse inquisitive men.

Once it is admitted that an intelligent being, human or Divine, may have tampered with evidence, that evidence is worthless. Fundamentally a person who is utterly convinced that special creation adequately explains the living world as we see it is in the same position as a scientist who does not trust a colleague. Both have explanations for the facts no matter what the evidence may seem to show; both go through meaningless motions if they bother to examine the evidence at all.

It is impossible to convert the extreme fundamentalist to a belief in evolution by any amount or type of evidence. The best that one can do with scientific evidence is to ask whether it is consistent with a given theory, whether it was predicted by the theory, and whether any part of it contradicts the theory. If all is consistent, we say that the theory is adequate to explain the original observation—in our case, the observed complexity of living things. And if our theory that all living things have evolved from earlier forms and that at progressively remote times there existed common ancestors to progressively dissimilar present-day organisms is adequate, we merely conclude that it is unnecessary to invoke special creation as a mechanism to account for the diversity of life about us. Furthermore, we reject special creation as an adequate explanation because we can think of no means by which we can put it to a valid test, because we can imagine no observation falling outside the capabilities of a Creator possessing unlimited ability. Search diligently for the adequate, reject the untestable—those are the recognized procedures of the laboratory, the classroom, the clinic, and the courtroom.

2

The Nature of
the Evidence

In mustering the support that genetics gives to evolutionary theory, a certain amount of technical information must be brought out and explained to the nonscientist. This will require the use of new words; at the same time, the reader will find it necessary to concentrate on unfamiliar ideas. It would be doubly difficult if we were to take up the nature of the experimental observations together with their interpretation, while struggling with the vocabulary itself. Fortunately, we need not undertake all of these tasks simultaneously. The nature of the genetic evidence or, rather, the logic behind what we will consider to be evidence can be illustrated by analogous arguments used in more familiar situations: court trials.

One type of evidence we will present later is very much like the evidence that a ballistics expert presents in a criminal trial. A murder has been committed. A suspect has been brought to trial. It is known that the suspect owns a certain gun. The state now attempts to prove that the murder bullet was actually fired from this gun. A ballistics expert is called into the case; the murder bullet is given to him, together with a second bullet known to have been fired from the gun in question. After an examination of the two bullets, the expert appears at the trial carrying photographs like the drawings in Figure 1a. The photographs and his observations form the basis of his testimony.

Precisely what does the ballistics expert contribute to a trial? What assumptions has he made? How is the reliability of his evidence to be judged? These are important questions. If we do not understand them, we have not really grasped his role. If we do not understand them, we, as jurors, will either accept his testimony on faith or reject it out of hand through ignorance; neither judgment satisfies our responsibility as members of a jury.

The milling machines that are used in rifling gun barrels leave a series of scratches on the interior surface of the barrel. These, in turn, score the lead slug as it is fired from the gun; the surface of a spent bullet is covered with a multitude of scratches. No one can predict in advance what pattern of scratches a given gun barrel will produce. The pattern is so complicated, however, that no two known gun barrels have ever caused precisely the same pattern. Each scratch can vary in respect to width, depth, and distance from its neighbors; the discernible patterns resulting from hundreds of scratches are virtually innumerable. Thus, when a ballistics expert testifies that a certain gun fired the murder bullet, he means that the pattern of scratches on the murder bullet coincides precisely with that on a bullet fired from a given gun. Furthermore, from experience he knows that the variety of patterns produced by different guns is so great that identical patterns are found only on bullets fired from the same gun. Consequently, having found the same pattern on two bullets, he testifies that the murder bullet was fired from the suspect's gun. (Had the lead slug not been available, the expert could have testified with equal assurance on the basis of scratches and other imperfections left on a spent cartridge found at the scene of the crime; each gun leaves its characteristic marks on the cartridge casings as well as on the lead slug, as Figure 1b shows.)

The reasoning employed by the ballistics expert is not restricted to the identification of murder weapons. A number of years ago a great deal of timber was stolen from a federal forest preserve. A federal agent, armed with cross-sectional slices from a number of stumps on the preserve, set out through the Midwest to examine lumber in lumberyards and on railroad cars. At last he found a shipment in Chicago in which the grain of certain boards matched the grain in the sections of wood he had brought from the federal land. This evidence led to the conviction of the thieves. The reasoning he used is precisely the same as the ballistics expert's. No one knows the exact pattern growth rings will take in a particular tree, but the patterns are so infinitely varied that two pieces of wood are unlikely to match unless they have come from neighboring segments of the same tree (see Fig. 1c).

A similar argument is used by scientists in dating certain ruins by examining growth rings in fragments of wood. In this case the argument says that wet and dry or cold and hot years impose an annual growth pattern on all trees in a given area. Hence, if fragments of two long-dead trees are found, one of which was old when the other was young, the growth rings representing the years when both were alive can be recognized and identified. Thus, the scientist acquires a pattern of growth rings extending from the innermost rings of the more ancient fragment to the outermost rings of the more recent. The search for overlapping periods of growth can be extended back in time by examining timbers from old fortresses and ancient dwellings. The pattern can be extended as long as one can find pieces of wood whose latest rings correspond with the earliest known rings of the reconstructed series. The procedure, although analogous to the two earlier ones, is not as precise since individual trees vary in their response to the same growth season; nevertheless, the procedure is both a sound and a scientifically profitable one.

We will find later that the hereditary material that is passed on so faithfully from parent to offspring also has its fine structure ("scratches," if you please) that can be identified under the microscope. Furthermore, we will see that identical fine-structure patterns arise only by duplication, and that identical fine-structure patterns are found in different species. Just as the federal agent argued that the lumber he discovered in Chicago was stolen originally from federal property thousands of miles away, so we will argue that the hereditary material identified by corresponding patterns of bands and threadlike lines, and found in different species, has in fact arisen from a common source.

Recent advances in modern biology permit us to carry this argument farther than we could have dreamed possible fifteen or twenty years ago. We now know that each of the thousands of different types of proteins—the complex material resembling egg white of which living matter is made—has a precise chemical structure. The building blocks of proteins are twenty relatively small molecules called amino acids (Fig. 2). Proteins themselves are enormous molecules. A molecule of human hemoglobin (hemoglobin is the protein in red blood cells responsible for the transportation of oxygen from the lungs to tissues in all parts of the body), for example, consists of four chains of 140 or more amino-acid molecules lined up in a very precise order (Fig. 3). The substitution of *one* amino acid for another in certain sites of these long chains can destroy the oxygen-carrying ability of hemoglobin; persons suffering from the hereditary diseases in which these substitutions are found generally die of severe anemias.

Right now we are interested in calculating the chance that a given sequence of amino acids will be found in two different types of protein molecules entirely by chance. In order to simplify our problem, we will ignore the apparent need to maintain certain amino acids in certain sites of the protein molecule; rather we will pretend that any one of the twenty known amino acids might easily be found at any one site. If, indeed, there are twenty amino acids available in equal frequencies, we can readily calculate that there is only 1 chance in 20 that the first amino acid in each of the two sequences will be the same, there will be only 1 chance in 400 that the first two amino acids of each of the two sequences will be identical, only 1 chance in 8000 that the first three and 1 chance in 160,000 that the first four amino acids of one sequence will match perfectly with the first four amino acids of another. Obviously the probability of precise correspondence decreases at an extremely rapid rate as the number of sites increases. Consequently, even though there are a few sites in any protein molecule at which there may not be complete freedom in respect to the amino acid to be found there, identical sequences of substantial numbers of amino acids in two different proteins (like the identical patterns of scratches on two different lead bullets) imply that they trace back to a common origin.

The reasoning followed above differs from that used in the discussion of scratches on bullets and growth rings of trees by virtue of our ability to specify an approximate probability that a given amino acid is to be found at a given site in a protein molecule (about 1 in 20 because there are twenty different amino acids). In this case our reasoning more nearly resembles that of a cardplayer who calculates the chances that a hand will contain certain cards. For example, ten cards are drawn at random from each of ten well-shuffled decks of cards and are placed face up in the same order in which they are drawn. What is the probability that one of the ten sequences will match another precisely? The chance that the first card in any one sequence is exactly the same as that in a second is 1 in 52, because there are fifty-two cards in each deck. Granted that the first cards in the two sequences are identical, there is 1 chance in 51 that the second will be identical because there are only fifty-one cards remaining. Consequently, the chance that one series of ten cards will be precisely the same as a second equals $1/52 \times 1/51 \times 1/50 \times 1/49 \times 1/48 \times 1/47 \times 1/46 \times 1/45 \times 1/44 \times 1/43$ or, roughly, $1/100,000,000,000,000,000$ (one in one hundred quadrillions). The probability that any one of the ten sequences of cards will match any other one is somewhat greater: there are $9 + 8 + 7 + 6 + 5 + 4 + 3 + 2 + 1$ or forty-five *different* pairs of sequences among ten; if any of these forty-five possible pairs were identical, we would say that a sequence had been duplicated. The chance for this duplication is still less than one in a quadrillion. Consequently, if such a duplication were found, there would be good reason to suspect that the duplicated series were in fact not independent, that the series drawn were not drawn at random. If we were shown not the cards themselves arranged in order but photographs of the cards, we would be justified in believing that the duplication represented two pictures of the same sequence. In this same spirit of skepticism, when we find that a given protein in each of two different species has virtually identical amino-acid sequences, or when we find that two different proteins found in individuals of any one species have virtually identical amino-acid sequences, we say that this is not a matter of chance; rather, we argue that the observed similarities in amino-acid sequences reflect common ancestral origins.

A final line of reasoning we intend to follow in presenting the case for evolution is beautifully simple. A fairly close analogy can be taken from tales of espionage. A man sits at a table in a European cafe; before him on the table are two fragments of a calling card. He is joined by a second man who takes a piece of cardboard from his pocket and places it between the two fragments on the table. The new piece fits perfectly; the calling card is now complete. Obviously, one card was torn into the three pieces which we now see assembled on the table. This simple analogy lacks the spirit of change that we will encounter later in discussing evolutionary genetics; it does not illustrate properly the predictions that a geneticist can make in the course of his studies.

A much better analogy can be found in the archeological reconstruction of ancient Sumerian writings. Imagine, for example, a Sumerian poet inscribing a dozen or more lines of poetry on a clay tablet. The inscription of these lines requires time (albeit a relatively short time) as the poet inscribes the first line, then the second line, and finally the last line. Even within a line a short time elapses while the poet inscribes it from one side to the other. Further, the poem itself represents the creative product of human thought; confronted with an incomplete tablet, modern scholars can make extremely accurate guesses about the words that were inscribed on the missing part.

Because of the many expeditions to the Euphrates Valley, fragments of Sumerian tablets are dispersed among universities and museums in a half dozen countries on at least three continents. Gradually the inscriptions on these fragments are being translated and the fragments are being reassembled into their original tablets. The evidence on which these reconstructions are based consists of the flow of ideas from the beginnings to the ends of interrupted lines, the flow of ideas from the beginning to the end of a poem, and the accuracy with which the pieces of clay themselves fit one another in jigsaw-puzzle fashion. It does not bother the student of Sumerian artifacts that the fragment bearing the center of a poem was found in Berlin while the remaining fragments were rediscovered in Philadelphia; if these widely separated fragments when reassembled reveal the consecutive thoughts of a creative poet, the reconstruction must be correct. It matters not that the fragments were finally located several thousand miles from one another, and five or ten thousand miles from their place of origin. We will see later that genetics furnishes us with equally beautiful examples. These are examples in which we will reconstruct the missing midportion of a sequence of events. And we will say that it matters not if our hypothetical reconstruction turns up unexpectedly in a second species; the sequence must have taken place in what was initially a *single* species.

3

Genetics: The Study of Heredity

Although we are concerned with evolution, our evidence is to be drawn largely from genetics. It is necessary, then, to acquire some familiarity with genetic terminology and a feel for the workings of heredity. Now is an appropriate time to point out, however, that this chapter is not meant to *teach* genetics; anyone wishing to learn genetics must find an appropriate text (a number are included in the section "References and Acknowledgments"). In subsequent pages, one will find a discussion of genetics adequate for an understanding of the arguments concerning evolution, though not for an understanding of the underlying genetic mechanisms.

Everyone is aware of the physical similarities of related individuals; these range from the startling likeness exhibited by identical twins to the ordinary person's possession of Aunt Emma's eyes, father's nose, and other features seemingly borrowed from sundry relatives.

One can speculate endlessly about the source of these similarities, suggesting that the cause is to be found in the mother's womb, in external influences such as chance visits by in-laws, or in mysterious fluids and vapors. Indeed, modern man has probably spent a good deal of his fifty thousand or more years speculating about this problem. The answer, however, was discovered only slightly more than one hundred years ago. In 1865 Gregor Mendel, after a prodigious series of carefully controlled breeding experiments using different varieties of peas, reported that characters appear among offspring in ratios that are predictable on the basis of certain simple assumptions. These assumptions—proved uncannily correct by later physical observations—were (1) that individuals carry two "factors" for each of many inherited traits; (2) that germ cells (sperm in males, eggs in females) contain only one of each pair of factors; (3) that germ cells of individuals who carry a dissimilar pair of factors are half of one type and half of the other (this is important; these germ cells do not carry some intermediate factor made by mixing the two original ones); and (4) that germ cells unite at random to give rise to the individuals of the next generation, individuals that carry *pairs* of factors once more.

A large number of ingenious experiments by many scientists, together with the development of excellent microscopical equipment, have permitted geneticists to extend Mendel's inferences and to identify the physical "home" of his hypothetical factors. Each feature of the abstract picture he developed entirely on the basis of mathematical ratios has been accounted for in the corresponding physical picture developed later. (Figure 4 represents these facts in diagrammatic form.) Each egg and sperm carries a number of microscopic bodies called *chromosomes;* the precise number of these is, with rare exceptions, constant within a species. At fertilization the newly formed individual comes to possess a *pair* of each chromosome. (If we regard the chromosomes of a single egg or sperm as the individual volumes of an encyclopedia, each with its own information, we see that the newly formed individual gets two copies of each volume, or two complete sets of the encyclopedia.) During the growth of the individual, there are numerous cell divisions. At each of these *all* chromosomes of the dividing cell are duplicated so that, following cell division, each new cell obtains its own double set of chromosomes. Only in the reproductive organs do cells undergo a special division that, in effect, *separates* the two sets of chromosomes. This special division accounts for eggs and sperm carrying only a single set of chromosomes.

The whole complicated procedure by which the germinal *reduction* division is accomplished defeats an otherwise inescapable truth: the individuals of any generation must have twice as many chromosomes as their parents. If this type of inheritance were to apply (and it would in the absence of a special reduction division), the number of chromosomes per cell would soon be astronomical; extinction of the species would be the prompt result of this uncontrolled increase of chromosomal material.

From the early 1920s through the 1940s the exciting problems of genetics were those concerning the relation of *genes* (as Mendel's factors are now called) to *chromosomes:* the tendency for certain genes to be inherited together in so-called *linkage groups,* the relation in both size and number of linkage groups to chromosomes (Fig. 5), the positions of genes relative to one another on the *linkage map,* the successful identification of specific genes with specific spots on individual chromosomes, and, finally, the identification of the chemical (DNA) that serves as the physical basis of heredity.

During this period, some biologists were busily describing the numbers and shapes and sizes of chromosomes in many different species of plants and animals. Entire chromosome complements were described, chromosome by chromosome, in great detail (Fig. 6). The organelle of the chromosome that governs its movement during cell division (the *centromere*) may lie near the end, in the center, or at some

16

intermediate spot on the chromosome; the position of this organelle determines whether the chromosome appears characteristically rod-shaped, V-shaped, or J-shaped during cell division. Chromosomes come in different sizes even within different cells of the same individual; these sizes were measured and recorded. Observations of this sort were so plentiful that atlases of chromosome numbers have been compiled.

A special word must be said in respect to (and out of respect for) the common vinegar fly, *Drosophila melanogaster,* and its relatives (Fig. 7). The vinegar fly was the most valuable experimental material of geneticists for decades. With the help of man's ingenuity, this fly has won two Nobel prizes: one for Professor T. H. Morgan in 1934 and another for Professor H. J. Muller in 1946. The solutions to most of the many genetic problems listed earlier were provided by experiments performed with the vinegar fly. An event of crucial importance to our arguments about evolution was the realization that certain enormous structures found within the cells of the salivary glands of *Drosophila* and other fly larvae (Fig. 8) are in fact chromosomes. The chromosomes of most cells are visible only when they condense into tightly coiled packets during cell division. For some unknown reason, the chromosomes in salivary-gland cells remain in an elongate condition, but then proceed to divide over and over again with the daughter strands remaining in close juxtaposition; thus, they become visible as enormously long structures. When a mature larva is ready to enter the pupal stage in preparation for the developmental transformations that change it into a winged adult, the total length of the chromosomal strands in a single salivary-gland cell is easily one millimeter (about $\frac{1}{25}$ of an inch). Through a good microscope the length of these chromosomal strands appears to be somewhat greater than a meter, about three or four feet. Throughout their length these chromosomes are marked with the most intricate series of wide, narrow, double, dotted, fuzzy, and sharp cross-striations. A great deal of the subsequent discussion will concern these

17

markings, which literally allow each chromosome to be mapped "inch-by-inch" along its length (Fig. 9).

The most recent and most exciting event in genetics pertinent to our story is the sudden and dramatic way in which scientists determined the physical structure of DNA, the chemical basis of heredity, and grasped the implications of this structure relative to self-duplication and the storage of information necessary for the operation of the biochemical factory of the living organism. Our earlier comparison of the chromosomal complement of an individual to the volumes of an encyclopedia was not altogether facetious.

Deoxyribonucleic acid (DNA) is a double-stranded chemical built much like a ladder which is twisted around its longitudinal axis so that its sides form a long double helix (Fig. 10). The helical twist, except to a physical chemist, is an incidental matter. To visualize the biologically important aspects of DNA imagine an ordinary ladder (Fig. 11) in which the sides are built of repetitive units (and, hence, can be made as long as one wishes) and are, let's say, twelve inches apart. Instead of being of one piece, each rung is built of two parts; some are built of four-inch and eight-inch pieces while the others contain three-inch and nine-inch pieces. Obviously, these 4-plus-8 and 3-plus-9 rungs can be arranged in any order along the length of the ladder; both pairs are twelve inches in combined length. Furthermore, a 4-plus-8 rung can be inserted in the ladder as either a 4-plus-8 or an 8-plus-4; the rung is still twelve inches long. The manner of insertion of the 3-plus-9 rungs is equally arbitrary.

Imagine now that we pull the two sides of the ladder apart and, in the process, the two pieces of each rung separate so that one piece goes with each of the two sides. Imagine, too, that we obtain two new sides (devoid of rungs) and an armful of assorted rung pieces and set about repairing each of the half ladders in our possession. We very quickly realize that we have absolutely no leeway in how we are to go about the job! Where we are confronted with an eight-inch stub of rung, we must attach a four-inch piece; a four-inch stub demands an eight-inch addition; a nine-inch stub demands three inches; and a three-inch demands nine. No other procedure will reconstruct the ladder. When we have finished repairing both halves, we find that the two completed ladders are identical and each is identical to the original one.

Here, then, is the bottommost level of reproductive processes among living things. The precision that is required at the level of molecular reproduction has been achieved by means of a double structure consisting of two complementary halves; when these are split apart, each half controls the rebuilding of its complement. The result is the formation of two identical double structures where there was one before. Indeed, now that we know the secret, we can ask, "Is there any simpler solution to the problem of reproduction?" I do not think that there is.

The preceding three paragraphs have dealt only with the accurate duplication of DNA. We said earlier, though, that DNA carries the directions for the biochemical factory of the living organism. Have we shed any light on this problem? Indeed we have. In describing our simplified ladder we pointed out that there was no restriction on the order of 4-plus-8, 8-plus-4, 3-plus-9, and 9-plus-3 rungs; each is twelve inches long and consequently fits between the two sidepieces regardless of the construction of its immediate neighbors. Let's look at one long sidepiece. Along its length we find four-inch, eight-inch, nine-inch, and three-inch pieces in any order. In effect, we have here a four-letter alphabet with which we can, with imagination, spell out messages at will. The English alphabet contains twenty-six letters but this is purely an accident; the Hebrew alphabet has only twenty-three characters, the Arabic has twenty-eight, and the phonetic alphabet now being tried experimentally in some schools contains forty-four characters. Even in the English alphabet the letter "double-u" suggests that the number twenty-six is not immutable. Four letters may not be very efficient for writing novels, to be sure, but we could manage. Morse code has (if we neglect the space) only two letters, dot and dash, but by assembling these into characteristic combinations we can transmit any written message, including novels (Fig. 12).

The four letters of the DNA alphabet are adequate for the job they perform, that is, spelling out the sequence of amino acids in protein molecules. The importance of this task cannot be overemphasized, nor should its difficulty be underestimated. Each protein consists of tens, hundreds, or even thousands of amino-acid molecules attached to one another in a specified order; slight alterations in the sequence in which amino acids are strung together can destroy the biological function of a protein molecule. It is now clear that the long ladderlike molecule of DNA serves as the blueprint for the construction of proteins, a self-replicating heritable blueprint. The four letters of the DNA alphabet are assembled in three-letter words along the length of the DNA molecule with one word standing for each amino acid. A point-by-point correspondence exists between the order of words on the DNA molecule and the order of amino acids in the protein whose synthesis it controls.

The relation between DNA and protein structures can be illustrated by the directions a chemistry professor might give his laboratory assistant. Imagine that models of the twenty amino acids are kept in large bins for classroom use. In order to demonstrate the structure of a certain protein, the professor writes the proper sequence of amino acids on a long paper tape and gives the tape to his assistant, who then connects the amino-acid models in the correct order.

The professor might write the following:

. . . Leucine.Histidine.Cysteine.Asparagine.Lysine.Leucine.Histidine. Valine . . .

On the other hand, there is no need for him to spell out each word in full, so he might prefer to write:

. . . Leu.His.CysH.Asp.Lys.Leu.His.Val . . .

This system follows the convention illustrated in Figure 2.

With a twenty-six-letter alphabet at his disposal, the professor might arrange with his assistant to let single letters stand for individual amino acids; twenty amino acids would require the letters from *a* through *t* (as in Figure 2). In this case the instructions might appear as follows:

$$\ldots e.h.o.s.f.e.h.c \ldots$$

Confronted, though with the four-letter alphabet of DNA, the professor would find that it is necessary to use three-letter words in order to instruct his assistant. Individual letters of a four-letter alphabet can be used to identify as many as four objects, no more. Combinations of two letters (two-letter words) of a four-letter alphabet can be used to identify as many as sixteen objects, not enough for twenty amino acids. But sixty-four three-letter combinations (words) can be made with the letters of a four-letter alphabet. These sixty-four combinations are sufficient to identify twenty amino acids (in fact, they allow for "synonyms," that is, two words having the same meaning). The professor, just as the living world of which he is a part, would be forced to use three-letter words from the DNA alphabet in order to write down directions for the construction of a protein molecule from twenty amino acids.

Some very complicated machinery lies between a length of DNA and the protein molecule whose synthesis it controls. This need not concern us any more than the gadgetry of an automatic typewriter need concern us when we say that a sequence of holes in a tape spells out the message the typewriter prints on a sheet of paper (or any more than the salary, terms of employment, or physical characteristics of the laboratory assistant in our analogy need concern us). There are twenty amino acids in naturally occurring proteins (Fig. 2); the position of each in its protein molecule is "written" by a three-letter sequence (a word) at a corresponding location on a DNA molecule.

4

Constancy and Variability of Chromosomes

This chapter will be devoted to a fuller discussion of chromosomes: their number, their appearance, and their fine structure. We will discuss their constant features; that there is a great deal of constancy has been suggested already. We will also discuss changes that appear from time to time in chromosome structure and, surprizingly, the subsequent constancy of these changes. By the end of this chapter we will have learned that chromosomal segments bearing identical banding patterns arise only by the duplication of earlier segments having identical patterns and, hence, trace ultimately to a common source.

Each individual starts his life when a sperm cell (contributed by his father) unites with an egg (contributed by his mother). Each germ cell has a single set of chromosomes, which we have already called a chromosome complement. The precise number of chromosomes and their appearance under the microscope depends upon the type of organism. Each germ cell of man carries twenty-three chromosomes, so that each individual receives forty-six chromosomes, twenty-three pairs. The array of chromosomes found in man is shown in Figure 13. The importance of the magic number forty-six, twenty-three pairs, for human beings is illustrated dramatically by the inconspicuous difference illustrated in (c) of the figure. There the "pair" numbered 21 has three, not two, chromosomes. The characteristic effect of this small imbalance is mongoloid idiocy; the person (male) whose chromosomes are depicted here, like all others suffering from mongolian idiocy, was mentally retarded. Imperfections in the precise assortment of any of the twenty-three chromosomes during germ-cell formation can have tragic effects.

During the life of an individual the chromosomes are duplicated faithfully at every cell division; each new cell gets two complete sets of chromosomes. A chromosomal analysis of any human cell—whether it is an embryonic cell, a bone-marrow cell, a circulating white-blood cell, or a loose cell scraped from the lining of the mouth—yields the same results: forty-six chromosomes that can be arranged as in Figure 13. In vinegar flies we always find chromosome complements of four pairs whether we examine the brain cells of larvae or the ovarian or testicular tissues of adults; the chromosomes shown in Figure 5, for instance, could have come from any of these sources.

The giant chromosomes of the salivary-gland cells of certain flies do not look like the chromosomes of the other body cells of these insects. The difference, however, lies primarily in their size. Gulliver, in his many travels, visited Brobdingnag, the land of the giants, where the English foot represented but a single inch. Under these circumstances, it may be recalled, Gulliver failed to appreciate the beauty of their women. Biologists, too, were unprepared to accept the beauty of giant chromosomes—indeed, they failed to identify them as chromosomes—when they were first discovered. An appreciation of the relative size of a chromosome as it appears in a salivary-gland cell and in a "normal" body cell can be obtained by examining Figure 9; while it requires an entire page to accommodate the map of the giant chromosome, a drawing (to the same scale) of the corresponding chromosome in other tissues would be no larger than any one of the heavy cross-striations shown in the detailed map. Compared to a giant chromosome, a regular chromosome is a mere dot.

In certain larger species of flies we can compare the giant chromosomes of a number of larval tissues. Even when examined in the minute detail permitted by these chromosomes, we see that the transmission of chromosomes from cell to cell during development is remarkably precise. Figure 14 shows giant chromosomes drawn from preparations of cells of the salivary glands and of the "kidney" (*Malpighian*) tubules of a Brazilian species of fungus flies. The salivary glands and Malpighian tubules are situated at widely separated places in the larval body. Large numbers of cell divisions intervene during the development of these two tissues from common ancestral embryonic cells. Nevertheless, chromosome duplication has proceeded so accurately that even the details revealed by the giant chromosomes have been preserved identically in the two tissues.

The differences that we do observe in the giant chromosomes of these two tissues concern enlarged regions, or "puffs." Now that geneticists have learned to appreciate these chromosomes, they realize that here lies one of the really exciting possibilities in the study of development. The puffed regions appear to be chromosomal regions actively engaged in synthesizing proteins needed for controlling the other activities of the cell at that moment. At any one time different tissues have different chromosomal puffs, while cells of one tissue will show different puffs at different times during the growth of the individual (Fig. 15).

Giant chromosomes let us discern rather easily the types of structural changes that can occur in chromosomes and the passage of these from parent to offspring. It is not obvious in Figure 9 but each of the giant "chromosomes" drawn there is actually *two* chromosomes, which are very tightly paired throughout their entire length. The pairing is so precise that it is difficult for even a trained observer to detect the two individual structures. But it can be done. One of the first and most avid students of salivary chromosomes, Dr. C. B. Bridges, was able to represent the paired strands separately; a small segment of one of his drawings has been copied for Figure 16. Here we can see that what appears to be a single structure is really two strands wrapped around each other. The bands are not continuous bands; they are formed by the pairing of corresponding bands in the two chromosomal strands.

Two features of chromosome pairing are important for what follows. First, the precision with which the individual bands of the two chromosomes match is striking. These bands pair more accurately, for example, than the individual grains in the blocks of wood shown in Figure 1. Remember that the two chromosomes have come from different individuals, the mother and the father of the dissected larva from which the preparation was made. Second, the close pairing of the two chromosomes in the salivary-gland cell is the result of a band-by-band attraction. The importance of this point will become apparent immediately.

Chromosomes are physical structures that can be broken. When they do break—either spontaneously (very rarely) or artificially under the influence of x-rays or other man-made radiation—the broken ends tend to rejoin. If the broken ends formed by a single break rejoin, it is impossible to tell that a break actually occurred. In this case geneticists say that the chromosome has "restituted"; it is whole once more. More interesting is the case in which the chromosome breaks twice; in this case one of the possible outcomes of "healing" is the inversion, end for end, of the chromosome piece lying between the breakage points. A chromosome carrying an inverted segment is sometimes referred to as an *inversion;* the term is not precise since only a segment of the chromosome is actually inverted. It is better to refer to the entire chromosome with its inverted segment as the "such-and-such" gene arrangement. Chromosomes with altered gene arrangements are transmitted from parent to offspring just as faithfully and accurately as are the chromosomes of the original gene arrangement from which they are derived.

The origin of an inversion through the breakage of a chromosome as we have just described it can be represented as follows (detailed diagrams of the origin of inversions are given, too, in Figures 17 and 18):

a. Original chromosome   A B C D E F G H I J K
b. Original chromosome
    with breaks (*)   A B*C D E F G*H I J K
c. Chromosome with inverted
    segment (breaks healed)   A B G F E D C H I J K

The giant chromosomes of flies allow us to recognize inversions easily and to map with extreme accuracy the breakage points that gave rise to them. Each giant chromosome, as we saw earlier, is really a pair of chromosomes joined band to corresponding band throughout their entire lengths. Suppose the larval fly carries one inverted and one noninverted (normal) chromosome, such as the two, *a* and *c,* represented above. The letters can represent bands; according to the rules of the pairing game A must join with A, B with B, C with C, and so on through K with K. To accomplish pairing of this sort when one chromosome carries an inverted segment, the chromosome must form a loop, as illustrated in Figure 18. There is no other way in which the pairing requirements can be met. True to this prediction, the giant chromosomes of fly larvae carrying chromosomes of different gene arrangements do indeed pair in this manner; several examples have been illustrated in Figure 19.

The pairing configuration exhibited by two chromosomes that differ by an inverted segment is important for the argument that the pairing of salivary chromosomes results from the mutual attraction of corresponding bands. Regardless of their physical location in respect to the tip and the base of the chromosome, corresponding bands of the giant chromosomes of fly larvae seek out one another and pair tightly with each other. To amplify the evidence in support of this contention, illustrations of the pairing of chromosomes carrying extra segments and small deletions are shown in Figure 20. When a larval fly carries two chromosomes, one of which lacks a small segment, the bands of the corresponding segment in the other chromosome have no pairing partners. And, indeed, they remain unpaired; under the microscope they are seen as a small loop projecting out from the giant chromosome. Bands of these giant chromosomes, in other words, pair only with corresponding bands; noncorresponding bands do not pair, even if their regular partners are missing. If instead of a deletion, one of the two chromosomes carries a small duplicated segment, the bands in this segment pair with the corresponding bands in both its own and the other chromosome. Under the microscope this segment of the giant chromosome appears even larger than usual since it contains three, rather than two, strands. It is amply clear, then, that corresponding bands, and only corresponding bands, pair in salivary-gland chromosomes. Corresponding bands, furthermore, are bands related by descent.

5

Reconstructing Inversion Sequences

The giant chromosomes of fly larvae serve as a keystone in our endeavor to show that species arise by evolution from other, earlier species. These chromosomes are so important because of the precision with which inverted segments in them can be described. Before proceeding farther, we must pause momentarily in order to emphasize facts that will be essential to an understanding of later arguments.

First, it is necessary to appreciate the variety of banding patterns that occur along the length of a giant chromosome. A student learns to identify the different chromosome arms in a given species by their obvious landmarks, usually tips and bases, within a matter of hours. A specialist learns to recognize landmarks at short intervals throughout the length of each chromosome; he frequently gives picturesque names to them, such as "shoe buckle," "goose neck," "duck's head," and "four brothers," just as a guide names landmarks in an unmapped wilderness. Finally, the expert who restricts himself to a small segment—for instance, in an effort to determine the precise location of a given gene (see, for example, Fig. 26)—may get to know each band in that segment so well that he can find it even when it moves to a new and unexpected location, as it might after an exposure to x-radiation.

Second, one must appreciate the enormous array of inversions that could be recognized visually (under the microscope, of course) if they occurred in one of the giant chromosomes. There are easily a thousand bands in the maps of the better-known giant chromosomes, such as those of the common vinegar fly. An experienced microscopist can easily locate a breakage point with an accuracy of two bands. Consequently, each break point of an inversion can be located at any one of some 500 spots at the very least. Since an inversion arises as the result of two breaks, 250,000 (that is, 500 × 500) different inversions could be recognized as different from one another, and each could be identified individually if the need arose. That is, a quarter of a million inversions could, if necessary, be identified by individual names. Each of these 250,000 gene arrangements could give rise in turn to another 250,000 two-break inversions, and these too could be identified. And so on. Fortunately, the need to use all of these identifying names has not arisen.

There is another use for these calculations: if two inversions occur independently in a chromosome of a certain gene sequence, there is less than 1 chance in 250,000 that they will be microscopically indistinguishable to an expert. Thus, if we find two chromosomes that differ in an identical manner from some arbitrary standard gene arrangement, and if the two match perfectly with one another (no inversion loops, no unpaired segments) when both are present in the same individual, we assume that these inverted chromosomes did not arise independently but, instead, are descended from an inversion that occurred once, and only once, at some earlier time. We assume, to repeat ourselves, that a chromosome with a given sequence of bands arises once and only once.

The distributions within the United States of several named chromosomal arrangements of one of our native fruit flies, *Drosophila pseudoobscura,* are shown in Figure 24. Even though these arrangements are found over areas covering tens of thousands of square miles, and even though some of them are found in disjunct areas separated by hundreds of miles, each named arrangement has arisen only once in the history of the species. Each has arisen at a given moment, in a given individual fly, and has spread throughout the areas it now occupies from that single source. In fact, some may have occupied much larger geographic areas in the past but have subsequently been replaced by other, more recent gene arrangements. On the other hand, other inversions have surely arisen in a similar manner, only to vanish without trace almost immediately.

Finally, in an effort to tie up loose ends, we must emphasize that inversions occur spontaneously extremely rarely. We emphasize this point in order to argue that any naturally occurring chromosomal arrangement arises from its predecessor by an inversion involving two, *and only two,* breaks. Literally, hundreds of thousands of salivary gland chromosomes have been examined in the laboratory. The number of previously unknown inversions that have been identified among those examined, inversions that are likely to have occurred in the laboratory, is no more than ten. The chance that a fly carrying chromosomes with identical gene arrangements will produce a germ cell carrying a rearranged version of this chromosome is probably no greater than 1 in 100,000, and probably much less. An approximately similar estimate is obtained by calculating the frequency with which a chromosome is expected to acquire coincident pairs of certain mutations. Mutations are heritable changes of the genetic material, some of which are caused by chromosomal breaks that have "healed" imperfectly.

This summary has been oversimplified, but we are not far from the truth when we say that a germ cell with three breaks of spontaneous origin will be some 300 times rarer than one with two breaks; one with four spontaneous breaks will be about 100,000 times rarer than one with two. Thus, it is far more likely that a newly arisen chromosomal rearrangement will differ from its ancestral arrangement by two breaks (the least possible number since one break cannot give rise to an inversion) than by three, four, or more.

On the basis of the above information, we are now able to make one of those simple claims that are so valuable to science (and so satisfying to scientists). We claim that the various gene arrangements found in natural populations of fruit flies (and other flies as well) have arisen one from the other in a sequence such that each differs from its predecessor by a simple inversion involving only two breakage points.

One consequence of this claim can be illustrated as shown in Figure 21. In this figure a chromosome, represented by the letters A through H, is arbitrarily broken between C and D and between G and H. After the inversion of the segment DEFG, the chromosome with the new gene arrangement is represented as ABCGFEDH. The newly rearranged chromosome is again broken in two places, between A and B and between F and E. After the inversion of the segment BCGF, the newest gene arrangement is represented as AFGCBEDH. Through two independently occurring inversions we have obtained chromosomes with three different gene arrangements; the process is repeated below in the form of a table (more or less repeating what is also presented in Figure 21):

Original gene arrangement   A B C D E F G H
Original gene arrangement, with breaks (*)   A B C*D E F G*H
New (second) gene arrangement   A B C G F E D H
Second gene arrangement, with breaks (*)   A*B C G F*E D H
New (third) gene arrangement   A F G C B E D H

(One word of explanation may forestall confusion in interpreting this sequence of events. The second gene arrangement arises in a single chromosome of the original arrangement, and it then spreads within the population of flies. The third gene arrangement arises in a single chromosome of the second arrangement. Chromosomes possessing the second arrangement do not disappear when the third arrangement arises.)

If we were to encounter the three gene arrangements ready-made in a wild species of flies, what could we say about the pattern of their origin? The stipulation that inversions arise only as the result of two-break arrangements compels us to reconstruct events in one of the following ways:

$$1st \longrightarrow 2d \longrightarrow 3d$$
$$1st \longleftarrow 2d \longleftarrow 3d$$
$$1st \longleftarrow 2d \longrightarrow 3d$$

The same stipulation eliminates the two possibilities, 1st $\longrightarrow$ 3d and 3d $\longrightarrow$ 1st, since these pairs of gene arrangements differ by four breakage points (between A and B, C and D, E and F, and G and H). Thus, we are able to form an orderly sequence of events to account for the origins of the various gene arrangements encountered in wild populations. (That we can interpret the origin of the three arrangements in any one of three ways, while only one of the three can be correct, is not nearly as important as the fact that we can bring *order* into the sequence.)

Sequences of gene arrangements have been discussed in purely theoretical terms so far, except for the illustration of three inversions—Arrowhead (AR), Pikes Peak (PP), and Standard (ST)—in Figure 19. These three arrangements form a sequence AR ↔ ST ↔ PP. In going from AR to PP (or vice versa) one must pass through ST because four breakage points differentiate the AR and PP gene arrangements. The entire sequence of gene arrangements found in *Drosophila pseudo-obscura* is illustrated in Figure 22. It is plain from this diagram that the AR–ST–PP trio forms only a small part of the entire sequence. It is impossible to illustrate all of the pairing configurations of the different gene arrangements in *Drosophila pseudoobscura.* Figure 23 shows, instead, the simple sequential patterns formed by four gene arrangements found in *Drosophila azteca,* a Mexican fruit fly; this figure shows the pairing configuration of every combination of two different gene arrangements.

Analyses of inversion configurations in different groups of flies confirm the points we emphasized at the start of this chapter. Whenever complex pairing configurations involving many chromosomal breaks have been encountered, a sequence of independent, two-break inversions has been found adequate to account for the derivation of one multibreak arrangement from the other. If four or more breaks occurred simultaneously (an event that would destroy the logic behind sequences of the sort illustrated in Figure 22), most of the possibilities arising from the rejoining of broken chromosome ends would be inexplicable on the basis of a series of simple, two-break events. A sequence not easily explained on the basis of a series of two-break events has yet to be found in natural populations.

In constructing the sequence of simple inversions that is necessary to explain observed complex-pairing configurations, one must at times postulate the existence of gene arrangements that have not actually been observed. (For example, if a geneticist studying the western fruit fly, *Drosophila pseudoobscura,* had observed only the last illustration in Figure 19, the AR–PP pairing configuration, he would have been forced to postulate the existence of ST and would have initiated a search for the AR–ST and ST–PP configurations illustrated as a and b of the figure.) In at least a half-dozen cases, gene arrangements that were predicted as necessary arrangements (hypothetical gene arrangements) have been discovered following their original description. This is a highly satisfying event in any science; that it happens in an analysis of evolutionary changes within a species is doubly satisfying.

As a final point, I want to emphasize an aspect of this type of analysis of chromosomal inversions that so far has been

33

left unmentioned. The sequences of gene arrangements we have been discussing with such care represent a series of changes that occur in the hereditary material of a species over a period of time. In none of our examples have we asked ourselves which sequence of genes (or, sequence of bands) in a chromosome is the primitive one for a species; instead we have been content merely to draw doubleheaded arrows between adjacent sequences. But, given any grounds for designating one of the chromosomal arrangements in a sequential diagram as the starting point, all arrows in the entire diagram immediately become singleheaded. Later on we will do just that; we will attempt to identify the starting (or primitive) gene arrangement. For the moment we will merely emphasize what may have been overlooked: the arrows in our diagrams represent both changes that have occurred in chromosomal structure and the relative temporal order in which these changes have occurred. These changes, occurring in an orderly sequence through time, are *evolutionary changes!*

6

Chromosomes in Related Species

The number of chromosomes possessed by individuals of a species is generally constant. Indeed, constancy is required if the offspring produced by the more-or-less random mating of male and female individuals are to be normal in respect to both individual well-being and fertility. It is inconceivable, for example, that some vinegar flies could have only one pair of large chromosomes, others two pairs, and still others three and four pairs. The sperm cells and eggs produced by these flies (if they existed) would carry one, two, three, or four chromosomes. Following random mating among these individuals, most offspring would not have *pairs* of chromosomes (the results of random mating in this type of population are illustrated in Figure 25); on the contrary, they would have any number, odd or even, from two to eight. Now the association of chromosomes as pairs is an essential maneuver in assuring the success of the reduction of chromosome number that occurs during germ-cell formation; it is the positioning of pairs on an equatorial plane that leads to the orderly passage of one chromosome of each pair to each of the two daughter cells. Individuals that do not possess pairs of chromosomes lack the basis for an orderly separation of chromosomes during the reduction division. These individuals in our hypothetical population would be virtually sterile.

Individuals that belong to different species need not possess the same number of chromosomes, nor need there be any obvious similarity in the morphological structure of the chromosomes they do possess. As a general rule, however, species that appear to be closely related do exhibit similarities in their chromosomal complements (Fig. 6). In groups that have been studied in great detail, it is possible to reconstruct the basis for different chromosome numbers. In some instances the differences in number to be accounted for are small, one or two; in others the differences in number are large, two-, four-, six-, or eightfold, for example. Cases of the latter sort are examples of what we will describe later as "instant speciation"; they stand in direct contradiction to Pope Pius XII's statement (p. 3) that "the process by which one species gives birth to another remains entirely impenetrable."

The observations to be discussed in this chapter (or, more precisely, the implications of these observations) will be clearer if we present a few details concerning the relation of genes (the units of heredity) and chromosomes (the physical bearers of genes). Earlier we were satisfied to say that genes are located on chromosomes; we did not commit ourselves to any greater extent even when we used the term "gene arrangement" synonomously with the observed sequence of bands in giant chromosomes. Now we want to be much more accurate. We want to make the point that genes are located at very specific points on chromosomes, that each gene has its own location which can be identified with pinpoint accuracy. Thus, genes—like bands—mark chromosomal segments related by descent.

The results of one experimental procedure for revealing the physical location of genes on chromosomes are shown in Figure 26. In an experiment of this sort, x-rayed normal male flies (the vinegar fly, *Drosophila melanogaster*) are mated to females that carry two genes which cause white eyes (no eye pigment). The normal gene for eye color (red eye) contributed by the male masks the mutant gene in the female offspring from this mating. That is, one expects, and generally observes, that the daughters produced by this particular mating are normal (red-eyed). Irradiation, however, causes an occasional loss of a chromosomal segment. The loss is explained by the occurrence of two breaks, a loss of the intermediate piece, and the reunion of the remaining broken ends, thus: ABC*D*EFGH → ABCEFGH plus D, a small fragment that is lost. If the deleted piece (D in the above example) includes the normal gene for red eye color, the daughter receiving the defective chromosome (ABCEFGH) will possess white eyes. These exceptional daughters are rare; nevertheless a number of chromosomes carrying such small deletions have been obtained and examined in salivary-gland cells. Since each deletion examined was known to remove the gene for red eye color, the location of this gene is found by identifying the chromosomal band(s) common to all missing segments. The results of this type of experiment are unsurpassed in the precision with which genes are located in chromosomes.

Experiments of the above sort have successfully revealed the location of a number of genes in the various chromosomes of the vinegar fly; other procedures have led to the approximate location of many more (see Fig. 5). The association of a gene with a specific band in the salivary chromosome persists even when the chromosomal complement is broken repeatedly by experimental means and the pieces are drastically reshuffled; it is the band itself, independent of its location among other bands, that carries the gene.

By deleting small chromosomal segments experimentally and by reshuffling the chromosomal complement by repeated breakage and rearrangement, we have concluded that a gene and the band at which it is located are inseparable. If we know the location of the band, we know the location of the gene; if we know the location of the gene, we know the location of the band. Suppose, then, that as a result of a comprehensive and comparative study involving a variety of *Drosophila* species (species as distinct from one another as house cats, bobcats, and lions are from one another), it appears that certain inherited abnormalities such as white eye color, vermillion eye color, missing bristles, and scalloped wing margins are common to all. It does seem, indeed, as if a great many *Drosophila* species fall heir to the same genetic ills. Can we take the above statement about genes and bands, whose validity within a species we have proved, and now apply it *between* species?

Before attempting to answer this question, it will be helpful to examine the chromosomes of different *Drosophila* species; in this way we may find at least a partial answer. The chromosomal complements of several *Drosophila* species are illustrated in Figure 27. Obviously the number of chromosomes in the different species is not a constant; *D. melanogaster* has four pairs, *D. pseudoobscura* has five, while *D. virilis* has six. Closer examination reveals, on the other hand, that each of these species possesses five long chromosomal arms and a "dot" chromosome; the long arms may be separate as they are in *D. virilis* or they may be joined to form large, V-shaped chromosomes.

The chromosome arms are the key to the gene-chromosome relationship that can be detected between species. There is, in fact, enough information regarding mutant genes to allow us to recognize groups of comparable mutants in different species and to relate these, in turn, to chromosomal arms (called "elements" in this type of study). There are six elements, each revealed by the mutant genes it carries. In *D. melanogaster* these are associated as follows to make four chromosomes: A, BC, DE, and F. In *D. pseudoobscura* the same elements (identified by the genes they carry) are associated in this manner: AD, B, C, E, and F. In *D. virilis* the elements are separate: A, B, C, D, E, and F. The *elements* and *chromosome arms* match *between* species in the same manner that *linkage groups* of genes and *chromosomes* match *within* species. Thus, we have strong presumptive evidence that there is more than mere chance involved in the construction of the chromosomal complements in different species of *Drosophila*. The chromosome arms appear to be related by virtue of their gene content. Relation in respect to gene content within a species means relation by descent; each gene, it will be recalled, is invariably associated with a given chromosomal band, and each band arises only by the duplication of a corresponding band in an ancestral chromosome. We conclude, therefore, that the corresponding elements of different species are related by descent.

Instant Speciation

The remainder of this chapter deals with still another aspect of chromosome number: specifically, differences in the numbers of chromosomes found in related species that appear to involve multiples of some basic number (twofold, fourfold, and eightfold differences are common). The situation we will reveal denies the statement that no one has yet succeeded in getting one species from another. On the contrary! If by a species we mean a group of organisms reproductively isolated from other groups so that its genetic future depends for the most part upon changes in its own genes, then we must admit that plants have a well-known technique by which they give rise to new species almost instantaneously. Some of these have arisen within historical times and, like *Spartina townsendi,* a European marsh grass, have been extremely successful. Man has made such plant species in an effort to improve the quality and quantity of his food supply. One such man-made plant—with the leafy top of a radish and the inedible bottom of a cabbage—has been singularly useless.

Plant species, to an extent much greater than animal species, frequently hybridize by mistake. The isolating mechanisms (procedures by which members of a species recognize and restrict their mating activities to those of their own kind) are much less absolute in the plant than in the animal kingdom. The reason probably rests in the complex nervous system of animals and the variety of recognition signals that animals can use in choosing a mate. Reproductive isolation of plant species, on the other hand, depends heavily upon such things as different flowering times, pollination preferences of insects, and habitat preference.

Once formed a hybrid plant is frequently a healthy, vigorous individual (in fact, so is the mule), but it generally produces a great deal of abnormal pollen and sets very few seeds. The chromosomes the hybrid has received from each parent form a complete set, so that the genetic directions for its growth and development are intact. The germ cells (eggs and sperm), on the other hand, receive only one chromosome of each pair (assuming that the chromosomes do pair in the hybrid; if they do not, the situation relative to fertility of the hybrid is worse than that to be described); most of these germ cells receive an assortment of maternal and paternal chromosomes. Now, if the chromosomes contributed by the two parents (members of different species, remember) do not carry precisely the same genetic information in each chromosome of the entire set, combinations of maternal and paternal chromosomes will not be genetically complete.

One can imagine as an analogy a pair of identical twins, each of whom has packed her entire wardrobe in seven different-sized suitcases. The sets of suitcases owned by the twins are identical. The two travel together on a European holiday, sharing outfits and combining items of apparel with complete abandon. This period of joint travel corresponds to the normal development of a hybrid plant. Finally, the twins decide to go separate ways. Each packs her belongings into her seven suitcases again. At the moment of parting, however, the hotel porter sends with each one a set of seven bags that is part hers, part her twin's. Unless they packed exactly the same types of clothing into each bag of the set, it is possible that the two set off, one with no outerclothing and the other with no underclothing. A similar, but less picturesque, difficulty faces germ cells in a hybrid plant.

Once in a great while, a cell will accidently fail to divide after its chromosomes have already duplicated. The result is a cell with twice as many chromosomes as a normal cell. Should this accident affect a cell early in the growth of a plant, an entire branch may be composed of such *polyploid* cells. In a normal individual, this doubling leads to irregularities in the formation of germ cells. The orderly reduction of chromosome number in preparation for germ-cell formation depends, as we have seen earlier, upon the presence of *pairs* of chromosomes. Members of each pair seek each other out, pair, and then pass separately into the two daughter cells. The presence of *four* chromosomes of each type in the polyploid cell leads to difficulties in the processes of pairing and orderly separation; groups of three or four come together, and then separation goes awry.

In contrast, hybrid plants that have undergone a similar accidental chromosome doubling are occasionally spared the difficulties arising from the presence of four identical chromosomes. The troubles besetting the fertility of some hybrids result, as we suggested above, from the almost total lack of pairing between chromosomes. Polyploidy, the doubling of all chromosomes, removes that source of trouble and—presto! —the hybrid is transformed into a true-breeding individual possessing one pair of every chromosome but with twice the number (or, more correctly, the sum of the two original numbers) of chromosome pairs of the two "parental" species. The new species is true-breeding; furthermore, it gives rise to infertile offspring when crossed with either parental species. (The mechanics of the procedure we have described here are not nearly as difficult as they may seem; the diagrams presented in Figure 28 may help to make them clearer.)

It is not our intent to catalogue the number of plant species of polyploid origin. We will merely assert that a great many plant species, cultivated and wild, are polyploids. Even polyploids hybridize, undergo chromosome doubling, and form higher "ploids." We will also assert that botanists have, by shrewd deductions, reconstructed the steps leading to the formation of certain naturally occurring polyploid species. They have identified the parental species and produced artificial hybrids in the laboratory. Finally, by the use of certain chemicals, they have doubled the chromosome numbers of the artificial hybrids. The experimentally produced polyploid hybrids look like their natural counterparts; furthermore many experimental and natural polyploids cross freely with one another and produce fertile offspring. There are no grounds for doubting that new plant species arise in nature precisely in the manner described here and precisely as the experimental botanist proceeds in his series of laboratory crosses. The contention (p. 3) that no one has as yet succeeded experimentally in getting one species from another, consequently, is wrong.

Animals cannot avail themselves easily of polyploidy as a means for forming new species; chromosome numbers in related animal species do not, as a rule, differ in simple multiples. Higher animals do not generally reproduce by self-fertilization or by asexual, clonal growth (as the growth of potatoes from tubers or of African violets from "cuttings"). In order for an animal polyploid species to arise successfully, the complicated combination of rare events—hybridization and doubling of chromosome number—would have to occur simultaneously in each of two individuals, one male and one female, living in the same neighborhood. Further, these two individuals would have to choose each other as mates in preference to normal individuals of the two parental species living in the same region. Their offspring, too, would have to prefer one another and mate, brother with sister, for a number of generations. For reasons that we need not describe here, the germ-cell formation in males and the sex-determining mechanisms would tend to malfunction. In brief, an animal species could adopt polyploidy only as a consequence of the coincidence of four, five, or six extremely rare events. When events are extremely rare, they coincide at infinitesimally small frequencies; coincident occurrences of several are rare even over time spans measured in millions of years.

7

Giant Chromosomes in Species Hybrids

The direction our main argument will take in coming chapters is rapidly emerging. We saw in the preceding chapter that within a species a certain small chromosomal region can be identified by the effect it has on normal development. For example, the compound eyes (the big eyes of the flies illustrated in Figure 7) of many *Drosophila* species are red. Once in a great while a vinegar fly with white (colorless) eyes is found; these mutant individuals are genetically defective. True-breeding stocks of white-eyed flies can be established and studied. Experiments of the sort described earlier (see Fig. 26) show that the gene whose malfunction causes white-eyed flies resides at a given band in one of the four chromosomes of vinegar fly. That is, in this band is found hereditary material normally responsible for producing red eyes, material which, if altered, causes pigment production to fail. Furthermore, if in a series of crosses white-eyed flies supposedly lacking this chromosomal segment should come to produce flies with red eyes, one can predict—and demonstrate under the microscope—that representatives of this particular band have found their way into the experimental stock of flies. Mutant genes serve as labels for particular bits of chromosomal material, for particular bands, and even for particular pieces of DNA.

We have also seen that a thorough analysis of many *Drosophila* species reveals that different species are commonly afflicted with comparable genetic abnormalities. In addition, when large numbers of these abnormalities are subjected to a detailed analysis, they are found to fall into similar clusters in different species (see Fig. 31). Finally, these clusters can be matched with the arms of the chromosomes that are identified microscopically. Having admitted earlier that mutant genes identify particular bits of genetic material, we can scarcely avoid concluding that the chromosomal material of different species is related. The entire pattern of association of the mutant genes in corresponding clusters suggests that all *Drosophila* species have arisen from a common source.

A third thread of our argument dealt with the pairing patterns of the giant chromosomes of the salivary-gland cells. In these chromosomes we saw the precise physical structure reflected in a most intricate pattern of light and dark bands (Fig. 16). We saw, too, the precision with which homologous chromosomes come together, the amazing accuracy with which these intricately marked chromosomes match up and pair with one another in flawless union. Furthermore, naturally occurring, as well as artificially produced, chromosomal rearrangements have shown us very clearly that it is indeed the attraction of identical bands which causes the pairing of giant chromosomes. Both the simple and the complex pairing configurations, configurations that can be predicted in advance, of salivary-gland chromosomes of different gene arrangements offer conclusive proof that the pairing attraction lies in the bands, not in the chromosome as a whole.

The ballistics expert examines his spent bullets, compares the patterns of scratches on each made by the gun from which it was fired, and decides whether the unknown gun that fired one was in fact the same gun that fired the other. Bands on chromosomes are to the student of evolution what scratches on bullets are to the ballistics expert. Identical patterns mean identical origins: each band arises only from a pre-existing band; a given pattern formed by hundreds of bands in a particular sequence arises only from a pre-existing, identical pattern. Large patterns that extend from one end of the chromosome to the other may be interrupted by chromosomal inversions, but break and rearrange the chromosomes as we will, the pieces with identical banding patterns seek each other out and unite, band by band.

What, then, do we find in the salivary-gland cells of hybrid larvae in those cases where different species of flies can be persuaded to mate? We see, first of all, evidence for tremendous differences between species in many (but not in all) such hybrids. But of greater importance by far, we see huge blocks of chromosomal material in which hundreds of bands are paired with one another, united as firmly and as accurately as the same bands are paired within individuals of a single species. To contend that these complex, identical patterns arose independently, that the observed identity of large chromosomal segments is a matter of chance alone, and that the pairing attraction has nothing to do with descent from a common origin is to turn our backs on every shred of logic, every argument, and every bit of evidence we have developed and presented so far. The simplest and most consistent explanation of the observed pairing patterns (examples of which are shown in Figure 29) is *common descent*. Identical segments have an identical point of origin; material once found only in a single species is now found in reproductively isolated species; two or more species, then, must have evolved from one.

The amount of chromosomal pairing in a hybrid fly is roughly, but not perfectly, correlated with the similarity of the parental species. The first illustration in Figure 29 shows the pairing pattern in one large chromosome of the hybrid between the vinegar fly (*Drosophila melanogaster*) and an almost indistinguishable relative (*Drosophila simulans*). Other than one large and several small inversion loops, loops of the sort found in the vinegar fly itself (but not identical to any of these), the banding pattern of the homologous chromosomes in the two species is virtually identical. Incidentally, these hybrid larvae are difficult to obtain; they develop into morphologically abnormal adults which are completely sterile. In other words, *Drosophila melanogaster* and *Drosophila simulans* are different species by any reasonable definition of the term.

In contrast to the nearly perfect pairing of chromosomes found in the two very similar species of vinegar flies, other hybrid *Drosophila* larvae can show chromosomal pairing that is limited to blocks of material here and there. The second example shown in Figure 29 is a hybrid between two western fruit-fly species (*Drosophila pseudoobscura* and *Drosophila miranda;* these species have no familiar names). The complexity of the banding pattern in each paired block is sufficiently great to rule out chance similarity. The unpaired chromosomal regions probably have pairing partners somewhere in the chromosomal complement, but the extensive rearrangement of chromosomal material that has taken place prevents all these segments from pairing successfully with one another.

Identical banding patterns of giant chromosomes have been used as evidence for genetic material that is related by descent from a common source; the argument is at its best in those cases where pairing can be observed in hybrid larvae. What can we do if two species cannot be crossed? Are we at a loss to draw any conclusions? No. A good observer can sometimes see similarities between two species, similarities that imply identity, merely by comparing the detailed chromosome maps of the two species. Figure 30 shows detailed chromosomal maps of *Drosophila pseudoobscura* and *Drosophila miranda,* the two species discussed above. Figure 30 also shows salivary-gland chromosome maps of a group of nonhybridizing *Drosophila* species from Latin America. In both instances one can detect landmarks and other points of similarity from an examination of the drawings alone.

We have not yet exhausted the evidence that one can muster to show that the genetic material of different species has come from a common source—in short, that evolution has occurred. We saw earlier that chromosomal rearrangements often permit us to reconstruct the sequence of inversions that have occurred within a species. We saw, for instance, in the now-familiar western fruit fly (*Drosophila pseudoobscura*) a collection of some seventeen gene arrangements that could be fitted into a sequence which reflects their origin, one from the other (Fig. 22). Such a sequence, constructed in a logical manner following the assumption that new rearrangements arise from earlier ones by two-break inversions only, represents an orderly series of events occurring in time. It represents temporal changes in the heredity material of a species. Hence, it represents evolution.

Hybrid larvae frequently have chromosomes that pair sufficiently well so that one can deduce the changes which have occurred in gene arrangements between as well as within species. The chromosomes of the two species of vinegar fly (Fig. 29), for example, differ by several inversion loops; identical loops do not exist within either species alone. It is interesting to note that the large inversion illustrated in Figure 29 was presumed to exist before it was actually observed under the microscope; its presence had been foretold from the contrasting patterns of association of corresponding mutant genes within the two species (Fig. 31). Again, we see that mutant genes do serve as precise labels for given segments of chromosomal material.

A sequence of gene arrangements especially valuable for tracing evolutionary changes from one species to another is the sequence which includes the inversions found in *Drosophila pseudoobscura* (Fig. 32 and p. 32), and others found in *Drosophila persimilis*. In the expanded sequence, we can see clearly the path of genetic change that encompasses two species. There is excellent evidence, too, regarding the relation of a third species as well. The diagram in Figure 32 follows the repatterning of genetic material from an arbitrary (but reasonable) starting point in one species through a point at which a second species branches off. From this point of branching, we follow the separately evolving inversion patterns of the two species. Not only do we claim in this case that the precise pairing of the chromosomes in the species hybrids shows that the chromosomal material has had a common source, but we also claim that the sequence of rearrangements that occurred in the chromosome reconstructs for us the precise pattern of change that led up to and then beyond the point of speciation.

The point made in the preceding paragraph deserves repetition: in confirming our belief that two chromosomes have descended from a common source, we can use not only the physical properties such as pairing attractions but also the patterns of change that lend themselves to sequential analysis. The first type of evidence, as we have pointed out before, corresponds to the evidence obtained by the ballistics expert. It is based on the observation that a band in a chromosome arises only from a pre-existing chromosome with the same band, and that a complex pattern of bands in sequence arises only from an earlier chromosome bearing the same sequence. The second type of evidence could be obtained in ballistics only if each bullet, as it was fired, made new imperfections in the gun barrel that would then appear for the first time only on the next bullet. In this case, not only could the bullets be identified as those which were fired from a given gun but they could also be arranged in the order in which they were fired. The latter information (which is sometimes available) adds a whole new area within which the ballistics expert can testify with authority.

The course of evolution has been traced above by the reconstruction of events that occurred in a step-by-step sequence through time. The final cases to be described here are based on the anomalous distribution of certain chromosomal inversions between species of *Drosophila*. These cases specifically involve rearrangements that permit us to say that a given sequence arose I → II → III, or III → II → I, or I ← II → III. These are the rearrangements that rule out (under the two-break-at-a-time rule) the possibility that I gave rise to III directly, or that III gave rise to I. There is, to be sure, a great deal of uncertainty associated with the directions the arrows should point, for we are seldom sure which of the three arrangements is the original one. We *are* sure, however, about the position of II. We know that II is the intermediate arrangement; either it arose from I and gave rise to III, or it arose from III and gave rise to I, or it gave rise to both I and III. The sequence does not fit any other possibility.

Overlapping inversions (those which permit us to construct the above sequences) found within a species demonstrate that the packaging of chromosomal material changes style periodically. Indeed, these inversions enable us to reconstruct the order in which the new-style packages arose. This reconstruction of events can be made, too, if arrangements I and II are found in one species while III is found in another. The simplest situation, many examples of which can be extracted from Figure 32, is the following one:

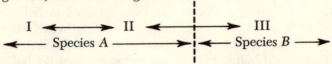

More spectacular, however, are the cases in which arrangements I and III are found in one species while II is found in another. This is a more spectacular finding for good reason: when he finds arrangements I and III in a certain species, the geneticist predicts the existence of an as-yet-undiscovered Hypothetical gene arrangement, II. He knows the breakage points of this arrangement as well as the pairing configurations it will make with the two known arrangements from the observations on arrangements I and III. It is exciting to discover a missing gene arrangement and to have predictions borne out; it is even more exciting to detect the postulated Hypothetical arrangement in a separate species. This situation can be represented in the following way (recall that HY, the hypothetical arrangement, is identical to II):

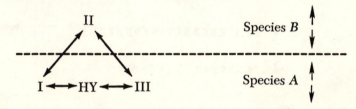

Since gene arrangement II falls between I and III no matter how one interprets the time sequence of these three gene arrangements, here is one of the strongest cases for the origin of species by evolution from earlier species. The entire sequence of three gene arrangements must at one time have been the property of one freely interbreeding population of individuals. Now we find the middle of the sequence in one species and the two ends in a second; furthermore, the division is maintained by all the isolating mechanisms by which species guarantee their genetic integrity. Such inversion sequences, shared by two species, serve as the fragmented Sumerian tablets of evolution.

51

8

Tracing Evolution Through Chromosomal Exchanges

Inversions—specifically, overlapping inversions—in the giant chromosomes of *Drosophila* flies are not the only changes in gene arrangement that reveal the order in which events have occurred. They are merely the ones that permit us to make the most refined analysis. An analogous series of changes occurs in many plant species. These are species that possess V-shaped chromosomes; in these species, chromosomes periodically exchange arms spontaneously. Although the chromosomes in these plants are small, exchanges between chromosomes have such striking effects that they can be studied with the microscope.

Our immediate job is to develop an understanding of chromosome exchanges and their consequences as far as microscopic examination of germ-cell formation is concerned. Consider a hypothetical plant species not unlike the evening primrose or the deadly nightshade. Plants belonging to this hypothetical species are composed of cells that contain seven pairs of chromosomes; each germ cell of these plants contains a set of seven chromosomes. Since each chromosome is V-shaped, there are fourteen arms in a set of seven chromosomes. Consequently, the set of chromosomes can be represented symbolically as:

$$1.2 \quad 3.4 \quad 5.6 \quad 7.8 \quad 9.10 \quad 11.12 \quad 13.14$$

Each pair of numbers connected by a dot represents a V-shaped chromosome; each number represents a particular arm.

An exchange of arms between two chromosomes of a set of seven can occur in a large number of different ways. As an illustration, let's assume that arm 4 changes place with 10 so that a new, or at least a repatterned, chromosome set arises.

$$1.2 \quad 3.10 \quad 5.6 \quad 7.8 \quad 9.4 \quad 11.12 \quad 13.14$$

Although germ cells carry only single sets of chromosomes, the plants themselves carry two sets. As these plants prepare to form eggs and pollen, the members of each chromosome pair come together in close association and then move apart at cell division so that one member of each pair goes to each daughter cell. This, we may recall, is the procedure by means of which the chromosomal content of a species is kept constant. Because chromosome arms 4 and 10 have changed places, both old-style and new-style chromosome sets exist in the plant species. Any one plant may have two "old," two "new," or one "old" and one "new" set. The pairing patterns that occur during germ-cell formation in these three types of plants are these:

TWO OLD SETS

| 1.2 | 3.4 | 5.6 | 7.8 | 9.10 | 11.12 | 13.14 |
| 1.2 | 3.4 | 5.6 | 7.8 | 9.10 | 11.12 | 13.14 |

TWO NEW SETS

| 1.2 | 3.10 | 5.6 | 7.8 | 9.4 | 11.12 | 13.14 |
| 1.2 | 3.10 | 5.6 | 7.8 | 9.4 | 11.12 | 13.14 |

ONE OLD SET AND ONE NEW SET

| 1.2 | 3.4 9.10 | 5.6 | 7.8 | 11.12 | 13.14 |
| 1.2 | 4.9 10.3 | 5.6 | 7.8 | 11.12 | 13.14 |

Plants carrying two old and two new sets of chromosomes are indistinguishable; each has seven chromosome pairs all looking very much alike. (These are not giant chromosomes; on the contrary, they are miserably small things to study.) Plants carrying one old and one new set, however, look different. Small as these chromosomes are, attraction is still between corresponding parts; hence, arm 4 of chromosome 3.4 attracts arm 4 of 4.9 while arm 9 of 4.9 attracts arm 9 of 9.10. The result of all this attraction is a circle of four chromosomes. This circle can be easily seen in material prepared for microscopic examination, such as is illustrated in Figure 33. (The notion of a circle was introduced into the above table by joining the terminal arms, No. 3's, by a line.) For simplicity, only two chromosome pairs are illustrated in the figure; an examination of the tabular representation given above reveals that the five chromosomes not involved in the exchange simply form ordinary chromosome pairs.

At some time subsequent to the first exchange, a second exchange of chromosome arms may take place in the new set of chromosomes. The second exchange may involve two entirely different chromosomes than were involved in the first exchange; for example, 1.2 and 13.14 may exchange arms to become 1.13 and 2.14. Or the second exchange may involve one chromosome that exchanged arms in the first exchange and a second chromosome that was not involved; thus, 3.10 and 7.8 may exchange arms to become 3.8 and 7.10. Finally, there is a very small possibility that the second exchange will involve precisely the same two chromosomes that were involved in the first one; if it did, the new chromosomes 4.9 and 10.3 would either give rise to 4.10 and 9.3 or revert once more to 3.4 and 9.10. The chromosome sets that would result from these subsequent exchanges (except for the rare case of involvement of precisely the same two chromosomes that exchanged in the first instance) can be described as follows:

a. OLD SET
1.2 3.4 5.6 7.8 9.10 11.12 13.14

b. NEW SET
1.2 3.10 5.6 7.8 9.4 11.12 13.14

c. SECOND EXCHANGE (INDEPENDENT OF THE FIRST)
1.13 3.10 5.6 7.8 9.4 11.12 2.14

d. SECOND EXCHANGE ("OVERLAPPING" THE FIRST)
1.2 3.8 5.6 7.10 9.4 11.12 13.14

Plants carrying two identical sets of chromosomes (sets that can be represented as *a/a, b/b, c/c,* or *d/d*) are indistinguishable even with the best microscope because the chromosomes and chromosome arms are so alike that 1.2, for example, cannot be distinguished from 1.13. Each type possesses seven pairs of chromosomes.

Individual plants can carry, however, any one of the six possible combinations of chromosome sets identified above as *a, b, c,* and *d.* The chromosomal pairing that would be seen in each of these combinations is shown in the following table:

```
a/b        1.2   3.4 9.10      5.6   7.8   11.12   13.14
           1.2  |4.9 10.3      5.6   7.8   11.12   13.14

a/c        1.2 14.13      3.4 9.10      5.6    7.8    11.12
          |2.14 13.1     |4.9 10.3      5.6    7.8    11.12

a/d        1.2   3.4 9.10 7.8      5.6   11.12   13.14
           1.2  |4.9 10.7 8.3      5.6   11.12   13.14

b/c        1.2 14.13      3.10   5.6   7.8   9.4   11.12
          |2.14 13.1      3.10   5.6   7.8   9.4   11.12

b/d        1.2   3.10 7.8      5.6   9.4   11.12   13.14
           1.2  |10.7 8.3      5.6   9.4   11.12   13.14

c/d        1.13 14.2      3.10 7.8      5.6   9.4   11.12
          |13.14 2.1     |10.7 8.3      5.6   9.4   11.12
```

This schematic representation of pairs of chromosome sets shows that exchanges of chromosome arms lead to circles of four, multiple circles of four, and circles of six chromosomes. Additional exchanges would, of course, lead to even more complex pairing patterns. Two independent exchanges give two circles of four chromosomes each; an "overlapping" exchange (one in which the same chromosome is involved twice) leads to a circle of six chromosomes. The chromosomes in these plant species are indistinguishable to the eye; nevertheless, the analysis of circles of chromosomes in combinations of chromosome sets systematically brought together as in our diagrams leads to a reconstruction of the sequence of individual exchanges. As in the case of inversions, it is necessary to assume that exchanges involve only two chromosomes at any one time. And, as in the case of inversions, the analysis leads to a reconstruction of a sequence of events (exchanges of chromosome arms) but does not in itself assign a direction to this sequence. (For example, the analysis does not tell us whether two chromosomes 1.2 and 3.4 gave rise through an exchange of arms to chromosomes 1.3 and 2.4, or whether the sequence of events was precisely the reverse: that 1.3   2.4 gave rise to 1.2   3.4.)

These details may not be tremendously exciting but they do show how a careful microscopic analysis and a careful and logical accounting for each ring of four, six, or more chromosomes lead inexorably to a reconstruction of the historical sequence of chromosomal exchanges. The result of such an analysis among plants of the nightshade family is shown in Figure 34. This "pedigree" is not well known, even among students of evolution. It is actually one of the most spectacular pedigrees available in the scientific literature because it traces the continuity of hereditary material through a total of ten species. This continuity makes no sense whatsoever unless we admit that the different species have evolved one from the other and, at some point in the distant past, from a common source.

9

Genes and Proteins

Every historical era has had its showcase of intellectual achievement—now in the humanities, now in the sciences, occasionally in both. During boom times for the humanities, the center of attraction may be art, music, literature, or one of the performing arts. Similarly, in science the baton of achievement has been at various times in the past in the hands of astronomers, of chemists, of physicists, or of biologists. Frequently we find biology at the fringes of what are striking advances in other branches of science; a case in point is the experimental synthesis of urea by Friedrich Wöhler in 1828. Until the moment of Wöhler's success, organic compounds were thought to be the product of processes occurring in living organisms, and in living organisms only. Of all organic compounds, urea (the excretion product of protein decomposition) is one of the most common, an excretory product of nearly every form of life. Suddenly a man showed that raw materials from the laboratory shelf could be united in the test tube to produce urea; at no time during the synthetic process was life involved.

The past decade has been one of frenzied advance in biology. The search for the chemical basis of heredity has ended. The chemical is DNA, deoxyribonucleic acid. Its structure and the features of its structure that made it suitable for the role it plays in the living world are known with virtual certainty. These features were described in Chapter 3. Our task here is to muster the evidence for evolution and the origin of species that molecular biology has given us. Molecular biology makes tremendous contributions to the study of evolution; to insure that these contributions will be properly appreciated, two small preliminary points will be described in some detail.

We have made the first point earlier, but we will repeat it. The four substances that form the cross-bars (the rungs) of the DNA ladder (see Figs. 10 and 11), by the order in which they occur, determine the precise chemical structure of protein. A single protein molecule (and all molecules of a given protein are identical) is composed of a large number of building blocks called amino acids. There are in living organisms twenty common amino acids. The sequence with which the various amino acids occur in a large protein molecule is determined by a corresponding series of cross-bars in the DNA molecule. A given combination of three cross-bars spells out a given amino acid just as certain combinations of holes in a punch-card spell out the words printed by a modern business machine.

The chemical analysis of proteins, difficult and laborious as it is, has proved to be an easier task than the sequential analysis of the cross-bars of DNA. However, if we know the code word in DNA that specifies which one of twenty amino acids will occur at a given site in a protein molecule, and if we know that the sequence of code words in DNA corresponds exactly to the sequence of amino acids in the protein, we can write down the structure of the DNA responsible for the synthesis of a given protein molecule. This can not be done now, but it will be done some day. Still, it is not necessary for our task. We are searching for evidence that the genetic material of two different species is related by descent. Our task is to show that large segments of DNA in two different species have an identical structure. To accomplish this task it is enough to learn that two protein molecules have identical (or largely identical) sequences of amino acids, for we know that the DNA responsible for the synthesis of two enormously complicated but identical protein molecules must also be identical.

The second point is a new one, and it requires that we return momentarily to the giant chromosomes that have served us so well in past chapters. Let's imagine a larval fly (I hate to say "maggot" for such useful creatures) whose chromosomes are entirely free of inverted gene sequences. The chromosomes in the salivary-gland cells of this larva pair tightly, band to band, throughout their length so that a microscopic examination reveals a simple configuration, devoid of complicating loops, such as that represented in the map in Figure 9. At least, this would be the type of picture seen in most cells and at most places throughout the length of each chromosome.

Here and there in cells where one expects to find this simple pairing configuration one encounters an abnormal pairing pattern—perhaps in a single cell in only one individual of many examined. The most common abnormalities are those in which certain adjacent bands pair not with their counterparts in the homologous chromosome but with what appear to be their counterparts in the same chromosome. By far the most frequent visible mispairing pattern is a chromosomal bulb in which there are two sets of nested U's, such as those shown in Figure 35. By lettering the bands as they appear along the chromosome, giving the same letter to bands that pair with one another, we reveal why the bands within these bulbs are sometimes called "reverse repeats": *a b c d d c b a*. During the process of pairing, these bands have two alternatives. They can pair with the corresponding bands in the homologous chromosome

$$-a-b-c-d-d-c-b-a-$$
$$-a-b-c-d-d-c-b-a-$$

or they can pair internally:

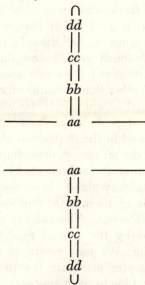

Stretched taut, the last configuration gives a bulb such as that represented in Figure 35. Similar bulbs were illustrated in Figure 30; there the bulbs served to identify corresponding chromosomal segments in four species of the *willistoni* group of *Drosophila*.

As nearly as one can tell, the relatively high frequency with which repeated chromosomal segments reveal themselves as bulbs can be ascribed to proximity. Pairing with neighboring bands, if these are proper pairing partners, must be about as simple as pairing with partners that are found on a different chromosome. Repeat segments, though, need not be physically adjacent in a chromosome; indeed, two repeat segments need not lie in the same chromosome. Under the pressures exerted by normal pairing forces, it is small wonder that the pairing of repeated segments such as those shown in Figure 35 occurs in only an occasional cell. Nevertheless, pairing of this sort must be recorded only once to show that the two chromosomal segments involved have arisen, during the course of the species's existence, from what was once a single segment. (The word "species" is used rather loosely in this sense; the bulb illustrated in Figure 30, for example, contains a chromosomal repeat that obviously arose before the four present-day *Drosophila* species diverged from one another.) This is the only known method, other than hybridization and the doubling of entire chromosomal complements, by which a species can acquire new genes. Repeats make "new" genes possible because the genes carried in the duplicated chromosomal pieces are largely free to take on new genetic functions.

The existence of chromosomal repeats has been known since the early days of salivary-gland chromosome analysis. To a large extent the role of these little duplications in evolution has been based on inference. The validity of these inferences has been buttressed by the refined analyses of amino-acid sequences in proteins. We now return to our main task by examining two examples involving the proteins hemoglobin and haptoglobin (a blood-serum protein).

Human blood, like that of all higher animals, owes its redness to hemoglobin, a protein constructed of four separate amino-acid chains together with four iron-binding groups. The four amino-acid chains of normal hemoglobin in adult human beings are of two sorts: two alpha chains and two beta chains. Before birth the developing embryo utilizes foetal rather than adult hemoglobin. Foetal hemoglobin is also made of four amino-acid chains: two alpha chains that are identical to those of adults and two gamma chains. As an added complication, normal adults carry a trace of still a different hemoglobin, one composed of two alpha chains and two delta chains.

No one knows precisely why man, during the course of his development, needs such a variety of hemoglobins. The variation in the second chain is not a matter of life and death for the individual (although it may affect his general health in a subtle way) because a few adults who have foetal rather than adult hemoglobin have been found. A large part of genetic research in the coming decades will be concerned with the identification of the biochemical control mechanisms which decide at a given moment in one's life that the manufacture of substance *A* will cease and the synthesis of substance *B* will begin. When this type of control is understood, we will be well on our way in understanding embryonic growth and development.

My enthusiasm for the problems connected with the control of human hemoglobin has sidetracked the discussion momentarily. What interests us at the moment is that each amino-acid chain—alpha, beta, gamma, and delta (see Fig. 36)—is made by a different gene. We are interested specifically with the evidence that these genes shed on evolution. And the point we will stress is this: it appears that all of the amino-acid chains used in the construction of hemoglobin are made by genes that have arisen by the repeated duplication of what was originally one gene, much as repeats in giant chromosomes have arisen by the duplication of genetic material. The basis for this statement is the great similarity in the amino-acid sequences one observes in these protein fragments.

The similarities of the alpha, beta, gamma, and delta chains can be easily seen in Figure 36. In man 141 amino acids strung together make an alpha chain, and 146 make a beta chain. The two chains can be placed side by side so that the same amino acid is found in each of 61 corresponding sites in the two molecules. Recalling that, with some limitation, any one of twenty amino acids may occupy a given site, we can see that identity at 61 of about 140 sites is much too much to be ascribed to chance. (Since the chance that two amino acids at a corresponding site is only 1 in 20, the chance that a run of six identical amino acids will occupy corresponding consecu-

tive sites in two molecules is about 1 in 64,000,000; the chance
that 61 of 140 amino acids can be matched up in two molecules
must be very small indeed.)

The beta and delta chains are even more similar than are
the alpha and beta; each contains 146 amino acids, and at most
only eight differ. The segments of DNA responsible for making
these protein fragments are virtually identical; furthermore,
a similarity so complex that it involves more than 400 rungs
on the DNA ladder (138 identical amino acids × 3 rungs per
amino acid = 414) virtually demands that the explanation lie
in the duplication of a small segment of genetic material.
Again, we find ourselves discussing a situation comparable
to the repeats we discussed earlier. The discussion can be
extended to include the beta and gamma chains (the beta of
adult and the gamma of foetal hemoglobin), both of which
have 146 amino acids, 107 sites of which are occupied by
identical amino acids. Once more the similarity is too great to
be brushed aside as chance similarity.

The story that we are developing here concerns the evolution of genes—the addition of new genes to the original set
through duplication and the subsequent alteration of these so
that they produce recognizably different proteins (or, more
exactly, chains of amino acids that serve as subunits for the
complex hemoglobin molecule). The evidence included repeat
segments in giant salivary chromosomes; these are segments
scattered here and there throughout the chromosome complement that occasionally pair intimately, thus revealing their
common origin. We have then used nearly identical amino
acid sequences as a means for recognizing genes of common
origin. The following account illustrates how the analysis of
proteins lets us reconstruct the past in even more detail.

A large number of proteins are found in human blood serum,
the fluid in which red and white blood cells are suspended. One
of these proteins is haptoglobin, a protein that apparently com-
bines with, and removes from circulation, hemoglobin that
has been accidentally released into the serum by disintegrating
red blood cells. Like hemoglobin, haptoglobin is composed of

two chains of amino acids, also designated as alpha and beta but unrelated, as far as is known, to the alpha and beta of hemoglobin. Different persons have slightly different haptoglobins. Persons can be identified as belonging to one or another of three categories that are designated as 1–1, 1–2, and 2–2; 1 and 2 can be interpreted both as alternative protein structures and as alternative states of the gene responsible for making this protein. A finer analysis reveals that 1–1 individuals can again be classified as belonging to one or the other of three types 1F–1F, 1F–1S, and 1S–1S where F and S refer to "fast" and "slow," an attribute important only in a certain experimental procedure. Once more, 1F and 1S can be regarded as both alternative protein structures and alternative states of the gene 1. It is not important for what follows, but you may want to confirm your understanding of this situation by noting that every person falls into one or the other of the following six classifications: IF–1F, 1F–1S, 1S–1S, IF–2, 1S–2, or 2–2.

The composition of haptoglobin molecules has been analyzed to some extent, though not as perfectly as that of the various hemoglobins. It is known that 1S and IF have very nearly identical structures: their lengths are identical and they have the same amino acids except for one (see Fig. 37). Among the amino acids that have been identified are the terminal and near-terminal ones. The amino-acid chain known as haptoglobin 2 is nearly twice as long as either 1S or 1F. In fact, it is formed by an almost perfect end-to-end attachment of these two molecules; at the point of juncture the last two amino acids of the leading half (1F) and the first two amino acids of the trailing half (1S) are missing. Of course the large protein molecule (haptoglobin 2) does not arise by the joining of the smaller protein molecules (1F and 1S); the duplication is a duplication of genetic material that controls protein synthesis. Our statement that a species can gain new genes by duplication of genetic material, a statement based on the occurrence of repeat pairing of salivary-gland chromosomes and supported by the resemblance of different hemoglobin molecules, appears to be confirmed. We cannot say when in human history the haptoglobin duplication occurred, but we know that there were already two forms of haptoglobin molecules, 1S and 1F. We know, too, that the duplication which gave rise to haptoglobin 2 occurred in an individual carrying both of these earlier genes. We do not know why a number of alternative genes should exist at any gene locus, but we do know that it is a common situation in human as well as in other animal populations; we say that these populations exhibit genetic polymorphisms. The analysis of the different haptoglobins has revealed the superposition of one polymorphic system upon a more ancient one during the descent of man.

An argument used throughout this book has been that two pieces of hereditary material, either a segment of a finely banded giant chromosome or a gene responsible for the synthesis of a complex sequence of amino acids, that are identical must have arisen from a single pre-existing piece. That is, a gene originally present only once in a chromosome complement comes, in one way or another, to be present twice. The two representatives then gradually diverge in the proteins whose synthesis they control, a divergence that reflects a gradual change in the DNA structure itself. The beta and delta chains of human hemoglobins are very similar; the segments of DNA responsible for their synthesis seem to be physically adjacent, as are many of the repeats in salivary chromosomes. Beta and gamma chains are somewhat less similar in chemical structure while the beta and alpha chains are even less alike. These resemblances and dissimilarities are interpreted as an illustration of the evolution of genes.

The above argument was restricted, as far as we have been concerned in this chapter, to changes, historical or evolutionary, that have occurred in the course of human evolution. The differences we remarked upon were differences between various and sundry amino-acid chains (proteins and pieces of proteins) found in and among human beings. We did not discuss, as well we might have, the time at which the gene duplications could have occurred. Are these changes a century old? Are they much older? Did they arise at the time of the ancient Egyptians? Or during the Ice Ages? These guesses are too short by far. These are trivial lengths of time as far as gene changes are concerned.

The gorilla has an adult hemoglobin which, like man's, consists of two alpha and two beta chains. The alpha chain of the gorilla differs from that of man by at most (and these have not been confirmed) two amino acids; the rest of the amino acids are identical. The beta chain of the gorilla differs from that of man by a single amino acid. Thus, we conclude that the two genes in the gorilla and in man responsible for the synthesis of the alpha and of the beta chains have descended from an identical initial pair. Because of the vast similarities between the alpha and beta chains in man, however, we decided earlier that these two genes themselves are the duplicated products of what was originally a single gene. Since the differences between alpha and the beta amino-acid sequences are relatively large (a difference at some eighty sites), and since both the gorilla and man have alpha and beta chains, we must conclude that the alpha-beta duplication antedated by an enormous period the evolutionary split that gave rise eventually to gorillas and to man. Indeed, if we pursue the matter further, we find that the alpha sequences of mice and men and the beta sequences of mice and men are more similar than are the alpha and beta sequences of either mice or men. Moreover, another protein, myoglobin, has been analyzed; there are obvious similarities between the amino-acid sequences of myoglobin from whales and hemoglobin from man. Thus, the proteins that are so useful to man seem to represent ancient discoveries; directions for their synthesis have been encoded in genetic material since long before many of the various groups of mammals split off from one another. Proteins of similar structure, like blocks of wood with similar grain patterns, tell us that the genes carried by many types of animals have had in primordial times a common source (Fig. 38).

10

The Universality of
the Genetic Code

Slowly, and with some repetition, we have developed a series of arguments, based upon genetical rather than anatomical or embryological evidence, in support of evolution. Most of our arguments have been based on the physical behavior of chromosomes. The precise and intimate pairing of homologous segments of giant chromosomes played an especially important part in these arguments. Patterns played their part, especially patterns of change in the arrangement of genes, patterns that lend themselves to sequential analysis, specifically the inversion of chromosomal segments and the reciprocal exchange of arms between different chromosomes.

Throughout we have avoided any substantial involvement in the details of genetics as a science. To learn a science (for example, to become a geneticist) is a difficult job; it requires a great deal of concentrated effort, years of effort. One cannot make a general case for evolution which demands that the reader first become a geneticist; the case will collapse unless the ordinary reader, preoccupied by other matters, can grasp the arguments without being bogged down in the technicalities of genetics itself.

While avoiding the minutiae of genetics, we have managed, nevertheless, to describe the physical basis of heredity. The steps leading to the successful identification of DNA as the bearer of hereditary information are as follows:

1. The recognition that certain bodies called chromosomes are inherited in precisely the same pattern as Mendel's hypothetical factors (genes).

2. The identification of specific chromosomes with certain heritable traits. First came the identification of the chromosomes responsible for sex determination; this was followed by the association of certain abnormalities with the inheritance of the sex chromosome. Later came the identification of clusters of genes and the successful identification of each of these "linkage groups" with a particular chromosome.

3. The localization of mutant traits at very precise spots on chromosomes (Fig. 26) and the demonstration that the physical sequences of genes on the chromosomes are identical to the sequences of genes in the clusters of genetically linked genes revealed by experimental crosses.

4. The demonstration that DNA, a chemical known for a long time to be a constituent of chromosomes, is sufficient in itself to transmit genetic traits. (This demonstration, before it was fully appreciated, had to await the development of a new branch of genetics, microbial genetics. Microbial genetics and its sister science, viral genetics, are the intellectual forerunners of today's molecular biology.)

5. The successful analysis of the physical structure of DNA, a structure we have discussed in some detail (Figs. 10 and 11).

Chromosomes, Giant Molecules, and Evolution

The structure of DNA is one obviously fitted for the role this chemical plays in heredity. First, it is a double structure so built that each half is a complement of the other. Split the halves apart and, in rebuilding two complete molecules, each half will specify that its new complement be an exact copy of the original. The two new molecules, then, will be identical. Second, the complete freedom of order of base pairs (the two-membered rungs of the spiral ladder) permits these to serve as a means for encoding information that can, at the proper time, be translated as a sequence of amino acids in protein molecules. These two properties are those which, at least with hindsight, we would ascribe to any substance that controls heredity.

Life depends upon reproduction. In everyday speech we sometimes speak of "division," as in "cell division," when we mean reproduction; this is a handy but inaccurate description. Division can proceed just so far before there is nothing more to divide. If one started with a large whale and each hour split it or its fragments in half, within three days the individual fragments would be as small as the smallest known viruses. Three days is a short time indeed. Reproduction proceeds by the building of new, complex chemical molecules. At the bottom of the reproductive process must be the faithful copying of the original set of directions. The problems of heredity are, in this sense, no different from those arising in the perpetuation of favorite family recipes.

We talk very casually about DNA. Perhaps we should pause and ask ourselves, "Are there any chemicals other than DNA that are used for hereditary purposes?" And if none other is used, isn't this a remarkable fact? It is indeed. Animals and plants, from the most complicated to the simplest, from the largest to the smallest, all utilize the same chemical for the physical control of heredity. Sorry, there is an exception: some plant viruses use another nucleic acid (RNA, ribonucleic acid) to transmit their hereditary properties. Since RNA happens to be an important intermediary between DNA and the synthesis of proteins in cells, this single exception does not represent a clean break with an otherwise general rule. (Superficially it appears that the "RNA viruses" have discovered a metabolic short cut which has eliminated their need for DNA; on the other hand, DNA may be a more recent "discovery" than RNA.)

The chemical used by nearly all organisms as the physical basis for hereditary information represents one aspect of universality; it is striking indeed that virtually all living things use DNA to determine heredity. But the question of universality can be raised, too, in respect to the genetic code itself: is the code that is used to spell out the amino-acid sequences in proteins universal? It need not be. In sending messages by telegraph, dots and dashes makes a convenient means of communication. Samuel F. B. Morse devised a code utilizing dots and dashes to indicate the different letters of the alphabet. The International Code, which has supplanted the Morse Code, also uses dots and dashes. Had it been known before the need arose that dots and dashes could be used in sending messages, any intelligent person could have composed a workable code. If several persons had worked independently of each other, there would have been as many codes as composers. Having been arrived at independently, identical combinations of dots and dashes would most likely have represented different letters in different codes. Doubtless, too, each code would have contained certain unique combinations of dots and dashes not included in any other (nonsense combinations as far as the others were concerned).

This analogy prompts us to ask, once more, about DNA: granted that virtually all forms of life use DNA to transmit genetic information, do they use the same code?

We cannot recount the story of the breaking of the genetic code in a detailed, comprehensible way, so we will answer the above question directly by saying, "There is only one known code." The type of experiment that supports this answer deserves brief mention. Fragmented bacteria, completely devoid of intact cells, are suspended in a mild preservative. Under these conditions the machinery of the bacterial cells still functions and proteins are still synthesized. The DNA in the suspension is then destroyed by the addition of a specific anti-DNA enzyme. When the DNA is destroyed there is no means by which RNA, the intermediary between DNA and protein synthesis, can be formed. Without this intermediary, all protein synthesis stops even though the raw materials—the twenty amino acids—are present. At this point the factory is intact but all operating instructions have been withdrawn; hence, the factory is idle.

With the entire bacterial factory intact and awaiting instructions, as it were, the molecular biologist can issue a number of instructions in order to see what the factory will do. For example, he may test RNA from the same type of bacteria as the stalled factory, but from a culture that is known to be making just one type of protein, an easily identifiable enzyme. Seeded with some of this RNA, the stalled bacterial factory in the test tube resumes making protein, but only the same enzyme that was being made in the culture from which the RNA came. This shows that the factory can be restarted after it has been brought to a halt.

Other tests involve the use of a series of artificially synthesized RNAs. Some of these are simple repetitive sequences involving only one of the four possible constituents of RNA. In response to each of these the bacterial factory produces a synthetic repetitive "protein" containing only one type of amino acid. More complicated tests follow in which the synthetic RNAs contain two of the four normal constituents. The frequencies of amino acids incorporated into each of the synthetic proteins are carefully measured and compared with the expected frequencies of various possible code words in the RNA. Gradually and systematically the code is broken. Man learns the language of genetics. More precisely, man learns the language of bacterial genetics.

The Universality of the Genetic Code

The final series of experiments consists of a repetition of those described but with other cells replacing bacteria as the source of the biochemical factory. Tests of this sort have been made on fragmented cells obtained from rabbits and rats, from normal tissues, from tumorous growths, from algae, and from plants. All cells tested so far have interpreted the synthetic RNAs in precisely the same manner. The code does indeed seem to be universal (Fig. 39).

The findings of brilliant studies such as these extend our earlier observations in a most dramatic way. Not only can we trace the historical changes in gene arrangement, not only can we work out the historical sequence of changing protein compositions, but we can now say (with only one or two exceptions) that DNA is the only hereditary material used by living organisms. And for its use, only a single code has been developed. Again, we find overwhelming evidence that life as we know it has evolved from a single initial source.

11

Summarizing the Case for Evolution

In the ten preceding chapters we have examined the evidence provided by genetic studies that leads us to conclude that life has had but a single source, and that it has evolved subsequently into what are recognized as the various forms of plants and animals we see around us. This evolution has been of two sorts: that which is represented by change alone, and that which has led to the formation of new species.

Nothing that we have seen contradicts or replaces the evidence that Darwin amassed in the preparation of his masterpiece, *On the Origin of Species*. Evidence gathered by geneticists supplements and enlarges that obtained by anatomists, embryologists, and taxonomists. The role of genetics in the study of evolution has been appreciated by geneticists for several decades.

The nonscientist and the beginning student, in contrast, generally encounters only the older types of evidence—evidence that failed to convince all persons when it was presented, evidence that still fails to convince all.

Students occasionally ask why chromosome morphology and chromosome pairing patterns should receive more consideration in tracing evolutionary progress than, let us say, some anatomical detail. Resorting to analogy, I would say for the same reason that recipes (directions written on scraps of paper) serve better in tracing the origins of regional cuisines than does the shade of the crust or the texture of a sauce. Chromosomes replicate; each chromosome arises as a copy of a pre-existing chromosome. Thus, chromosomes carried by today's organisms—by each individual organism—are the end products of long lines of descent. We have endeavored in this book to show that these lines converge as one goes back in time. Anatomical features, on the other hand, do not replicate; there are no lines of descent for color patterns of wings, or for muscles of the leg. Anatomical features reveal what a given set of genes can do under given environmental conditions; generally speaking, closely related sets of genes do remarkably similar things, and so small morphological differences can be used to trace evolutionary patterns. The correspondence between gene endowment and external features, however, is by no means absolute; there are a myriad gene combinations capable of producing the final form and substance of any living thing.

These chapters, however, were not written for the biologist alone; they were written as well for those who in one way or another are involved in the problem of teaching or not teaching evolution in our public schools. Included among these, of course, are the biology teachers; although the material presented here will not be new to them, perhaps it will be useful nonetheless. Also included among these persons are clergymen —clergymen whose colleagues are, to no small extent, the main source of arguments against the teaching of evolution.

And, finally, parents must be included among those for whom this book was written; some of these parents are quite likely to be called upon to vote on the continued teaching of evolution in their local schools.

At the very outset the following point was conceded: any person who is firmly and unalterably convinced that each of today's species of plants and animals arose by an act of special creation will find no evidence in this book that will compel him to change his mind. There simply is no such evidence, nor can there ever be. A Divine Being of infinite wisdom, we must all admit, could have created living forms in a manner that would have dribbled off as by-products all of these things we have gleaned as evidence for evolution. We can only say that He went about His task in a way that mimicked evolution in every detail; it is unfortunate that some event did not occur which would have clearly ruled out evolutionary theory.

Persons maintaining strong convictions against evolution despite the mass of favorable evidence should be aware, however, that for them all seemingly logical phenomena are of questionable value. These phenomena include evidence brought forth in court trials (such as that discussed in Chapter 1 and illustrated in Figure 1), as well as everyday decisions carefully arrived at by weighing observations in an effort to reach a sound judgment. There is no rational basis for accepting evidence as compelling in one circumstance while rejecting equally good or better evidence in a second, perhaps emotionally charged, circumstance.

While evidence of the sort we have summarized in the preceding chapters may not convince the emotionally committed persons, it may sway those with more open minds. The procedures of science are too modest to overwhelm blind faith; they are designed to influence the reasonable. Science does not construct explanations that can account for all conceivable observations and then seek in vain to find nonexistent contradictory evidence. Science cannot afford such a luxury; we would be back in the witch doctor's tent or the fortune-teller's booth in a remarkably short time if we were to embark on such a procedure.

Science can at best propose explanations for observations. These explanations, if they are fruitful, lead to certain expectations. They can be tested. Sometimes it happens that certain expectations follow equally well from each of two dissimilar explanations. If the two are really dissimilar, a situation will arise (usually after hard thought on the part of someone) in which one but not the other will account for a new set of observations. These are the forks in the road, the points of decision, that mark the advance of science.

Thus it is with the theory of evolution. Rejecting as we must the untestable "explanation" of the origin of species through special creation and proposing instead (1) that freely breeding populations change (evolve) with time and (2) that reproductively isolated populations of individuals (species) arise from similar but pre-existing populations, we have proceeded to itemize that evidence from the science of genetics we considered pertinent. A century ago, in undertaking a similar task, Darwin had no alternative but to examine the geologic record, to describe paleontological finds, to study geographic distributions of related forms, and to muster the evidence provided by anatomy and embryology. The case Darwin built still stands. Here we have supplemented his observations with those drawn from genetics, cytogenetics (genetics supplemented by microscopic studies), and molecular biology. Darwin, as it were, rounded up witnesses, checked alibis, got descriptions, and established the order in which events occurred; we have checked fingerprints, blood samples, and powder burns. There is no need to discard Darwin's theory of evolution. On the contrary, as one geneticist wrote in response to the statement of Pope Pius XII: "The occurrence of evolution in the history of the earth is well established as can be any event or process not witnessed by human observers, not witnessed for the simple reason that such observers did not yet exist or did not know how to record their testimony."

Figures

FIGURE

1

Evidence commonly accepted as conclusive in deciding whether two events are related or are independent.

(a) Markings left on two bullets by imperfections within the gun barrels from which they were fired. The markings shown here coincide perfectly, scratch by scratch; this is proof that the two bullets were in fact fired from the same gun.

(b) Firing-pin indentations of two spent cartridge cases. Again, the identity of the two sets of markings tells us that the same firing pin (and, hence, the same gun) was used in firing these cartridges.

(c) Blocks of wood whose grain patterns either match (the outside pairs) or fail to match (the innermost pair). The blocks with matching patterns were cut from corresponding segments of the same piece of lumber.

(a, b): Based on photographs made available through the courtesy of J. Edgar Hoover, Director of the Federal Bureau of Investigation.

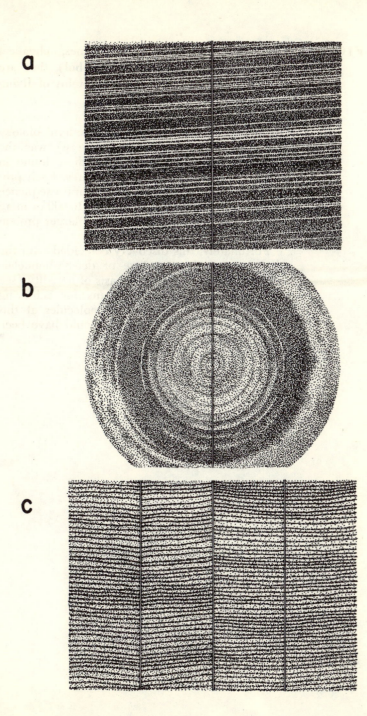

FIGURE

2

The twenty amino acids (names, chemical structure, and conventional symbol) that are essential for the synthesis of proteins of living organisms.

An outstanding discovery of modern biology (perhaps *the* outstanding discovery) was the revelation that proteins such as those found in meat or eggs have precise structures. Each protein consists of a unique and accurate sequence of these amino acids—as many as 10,000 or more amino-acid building blocks in the larger protein molecules.

Fortunately, this discovery coincided with the discovery of the structure of deoxyribonucleic acid (DNA), the chemical basis of heredity (see Fig. 10); otherwise, the means for achieving precision in the synthesis of molecules at this level of structural complexity would have been a baffling mystery.

a. $H_3\overset{+}{N}-CH_2-COO^-$
Glycine (GLY)

b. $H_3\overset{+}{N}-CH-COO^-$
  |
  CH_3
Alanine (ALA)

c. $H_3\overset{+}{N}-CH-COO^-$
  |
  CH
  / \
  CH_3 CH_3
Valine (VAL)

d. $H_3\overset{+}{N}-CH-COO^-$
  |
  $HC-CH_3$
  |
  CH_2
  |
  CH_3
Isoleucine (ILEU)

e. $H_3\overset{+}{N}-CH-COO^-$
  |
  CH_2
  |
  CH
  / \
  CH_3 CH_3
Leucine (LEU)

f. $H_3\overset{+}{N}-CH-COO^-$
  |
  CH_2
  |
  CH_2
  |
  CH_2
  |
  CH_2
  |
  $\overset{+}{N}H_3$
Lysine (LYS)

g. $H_3\overset{+}{N}-CH-COO^-$
  |
  CH_2
  |
  CH_2
  |
  CH_2
  |
  NH
  |
  C
  / \
  H_2N $\overset{+}{N}H_2$
Arginine (ARG)

h. $H_3\overset{+}{N}-CH-COO^-$
  |
  CH_2
  |
  $C=CH$
  | |
  HN $\overset{+}{N}H$
   \ /
    C
    |
    H
Histidine (HIS)

i. $H_2\overset{+}{N}-CH-COO^-$
  | |
  CH_2 CH_2
    \   /
    CH_2
Proline (PRO)

j. $H_3\overset{+}{N}-CH-COO^-$
  |
  CH_2
  |
  OH
Serine (SER)

k. $H_3\overset{+}{N}-CH-COO^-$
  |
  CH
  / \
  CH_3 OH
Threonine (THR)

l. $H_3\overset{+}{N}-CH-COO^-$
  |
  CH_2
  |
  COO^-
Aspartic Acid (ASP)

m. $H_3\overset{+}{N}-CH-COO^-$
  |
  CH_2
  |
  CH_2
  |
  COO^-
Glutamic Acid (GLU)

n. $H_3\overset{+}{N}-CH-COO^-$
  |
  CH_2
  |
  C
  / \
  HC CH
  || ||
  HC CH
  \   /
   C
   |
   OH
Tyrosine (TYR)

o. $H_3\overset{+}{N}-CH-COO^-$
  |
  CH_2
  |
  SH
Cysteine (CYS)

p. $H_3\overset{+}{N}-CH-COO^-$
  |
  CH_2
  |
  CH_2
  |
  S
  |
  CH_3
Methionine (MET)

q. $H_3\overset{+}{N}-CH-COO^-$
  |
  CH_2
  |
  C
  / \
  HC CH
  || ||
  HC CH
  \   /
   C
   |
   H
Phenylalanine (PHE)

r. $H_3\overset{+}{N}-CH-COO^-$
  |
  CH_2
  |
  $C-C-CH$
  || || ||
  HC CH
   \ / \ /
   N C
   | |
   H H
Tryptophane (TRY)

s. $H_3\overset{+}{N}-CH-COO^-$
  |
  CH_2
  |
  C
  / \
  O NH_2
Asparagine (ASP·N)

t. $H_3\overset{+}{N}-CH-COO^-$
  |
  CH_2
  |
  CH_2
  |
  C
  / \
  O NH_2
Glutamine (GLU·N)

FIGURE

3

An illustration of the manner in which amino-acid molecules are joined together in the formation of proteins.

The protein illustrated here is beef ribonuclease, an enzyme easily obtained from the pancreatic glands (sweetbreads) of cattle. Six of a total of 124 amino acids are shown in chemical symbols (above) together with the more convenient shorthand (below) illustrated in Figure 2.

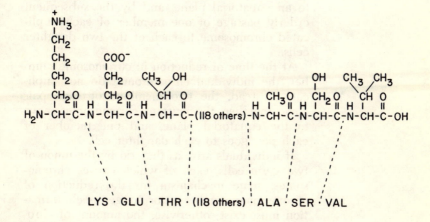

FIGURE

4

The contrast in methods of chromosomal apportionment in an ordinary cell division (right) and in the special reduction divison (left) that must precede germ-cell formation.

In an ordinary cell division each daughter cell acquires two complete sets of chromosomes identical to those of the parent cell. This is accomplished by the duplication of each chromosome, by the movement of all duplicated chromosomes to an equatorial plane, and by the subsequent orderly passage of one member of each duplicated chromosome to each of the two daughter cells.

At the time of reduction in chromosome number, the individual chromosomes do not duplicate. Instead, the two members of each pair come together, the *pairs* of chromosomes move to the equatorial plane, and one member of each *pair* goes to each daughter cell.

If individuals arise as they do by the union of two germ cells, each of which carries chromosomes, some mechanism for the reduction of chromosome number prior to germ-cell formation must exist; otherwise, the amount of chromosomal material per cell would double each generation. This type of chromosomal doubling would quickly lead to the extinction of the species.

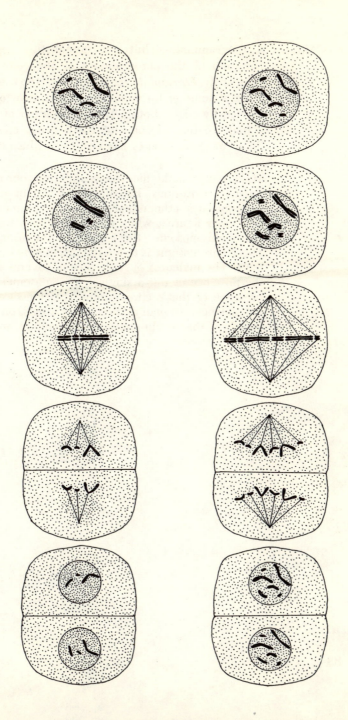

FIGURE

5

A diagrammatic (but not unrealistic) representation of the chromosomes of the common vinegar fly, *Drosophila melanogaster* (top), with a list of some of the mutant genes known to be carried by each (bottom). An indication of the actual location of some of these genes on the different chromosomes has also been illustrated.

Male and female flies have different chromosomal constitutions. Females carry a pair of X chromosomes (the rod-shaped chromosomes in the upper figure), while males carry one X and one Y chromosome (the J-shaped chromosome). The Y chromosome is to a large extent genetically inert. The maleness of male flies is governed by the presence of but a single X chromosome; the presence of the Y chromosome is not necessary for maleness. In man and other mammals, on the contrary, the Y chromosome determines maleness.

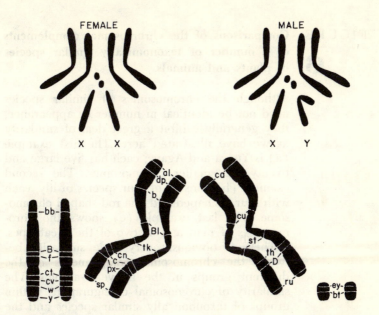

FEMALE   MALE

X X   X Y

CHROMOSOME I or X	CHROMOSOME II	CHROMOSOME III	CHROMOSOME IV
bar eye (B)	arc wing	bithorax	bent wing (bt)
bobbed bristles (bb)	aristaless (al)	blistery wing	eyeless (ey)
carmine eye	black body (b)	cardinal eye	grooveless scutellum
crossveinless wing(cv)	blistered wing	claret eye (ca)	shaven bristles
cut wing (ct)	bristle (Bl)	crumpled wing	
dusky wing	brown eye	curled wing (cu)	
folded wing	cinnabar eye (cn)	deformed eye	
forked bristles (f)	curved wing (c)	delta veins	
hairy wing	dumpy wing (dp)	dichaete wing (D)	
lozenge eye	expanded wing	divergent wing	
miniature wing	gull wing	ebony body	
prune eye	jammed wing	glass eye	
raspberry eye	jaunty wing	glued eye	
roughex eye	lanceolate wing	hairless	
sable body	light eye	javelin bristles	
scute bristles	lobe eye	maroon eye	
scalloped wing	narrow wing	pink eye	
silver body	net veins	prickly bristles	
singed bristles	plexus veins (px)	rotated abdomen	
tan body	purple eye	rough eye	
vermilion eye	reduced bristles	roughoid eye (ru)	
white eye (w)	rolled wing	scarlet eye (st)	
yellow body (y)	speck wing (sp)	sepia eye	
	star eye	stripe body	
	straw body	stubble bristles	
	thick legs (tk)	thread aristae (th)	
	vestigial wing	tilt wing	
		veinlet	

Comparisons of the chromosomal complements of a number of taxonomically similar species of plants and animals.

Although the chromosomes of similar species need not be identical in number or appearance, they generally exhibit a great deal of similarity as we have illustrated here. The first example (a) is Yucca and Agave; each has five large and twenty-five smaller chromosomes. The second example (b) consists of four species of lily, each with four V-shaped and one rod-shaped chromosome. The last example (c) shows the chromosomes of man and of two of the great apes.

There is obviously very little similarity between the chromosomal complements of the different groups in the above examples. The similarity of chromosomal configurations within groups of taxonomically similar species and the lack of similarity between such groups is not compelling evidence, but it suggests that taxonomically similar species may in fact share chromosomes drawn ultimately from a single source.

(a): After McKelvey and Sax; courtesy of the Arnold Arboretum. (b): After Warmke; courtesy of the *American Journal of Botany*. (c): After Chiavelli; courtesy of *Caryologia*.

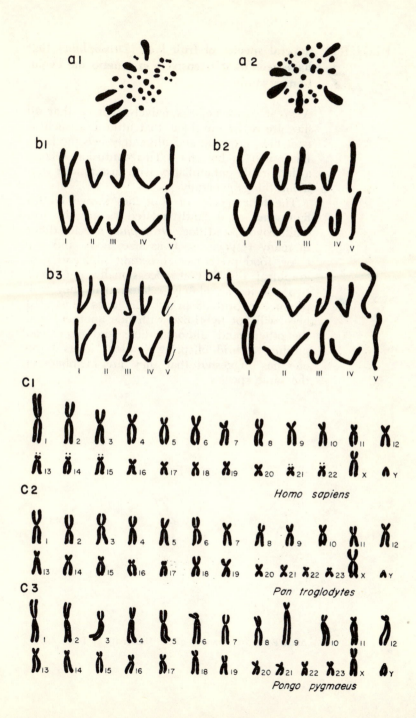

a1   a2

b1   b2

I II III IV V   I II III IV V

b3   b4

I II III IV V   I II III IV V

C1

Homo sapiens

C2

Pan troglodytes

C3

Pongo pygmaeus

FIGURE **Several species of fruit flies (*Drosophila*) that**
7 **have been used extensively in genetic and evolu-**
tionary studies.

A great many persons naively believe that all
flies are really the same, that little flies, such as
fruit flies, grow up and ultimately become large
flies, such as horseflies. This confusion reflects
a deplorable unfamiliarity with insects and with
living things generally.

The four species of fruit flies shown in this
figure are as distantly related as are the fox,
wolf, coyote, and dog. *Drosophila* species differ
in many respects, such as size, body and eye
color, food preference, breeding site, and geographical distribution. Occasionally one finds
two species so closely related that they will
produce hybrid offspring (the ones illustrated
here will not hybridize with one another). On
the other hand, although lions and tigers can
produce hybrid offspring (called "ligers"), no
one has suggested that they are members of
the same species.

After Patterson; courtesy of the University of Texas.

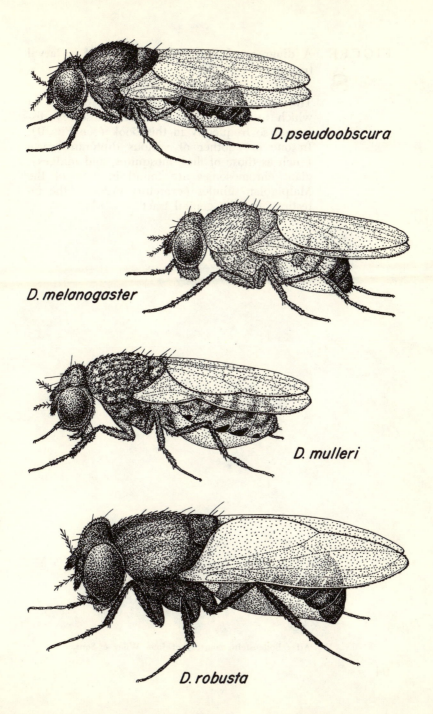

D. pseudoobscura

D. melanogaster

D. mulleri

D. robusta

FIGURE

8

A diagram of the intestinal tract of a larval fruit fly.

The salivary glands are the source of cells in which are found the giant chromosomes discussed so frequently in this book (see Fig. 9). In one or another of various dipteran larvae (such as those of flies, mosquitos, and midges), giant chromosomes are found in cells of the Malpighian tubules (excretory organs), the intestine, and the genital tract.

After Bodenstein; courtesy of John Wiley & Sons.

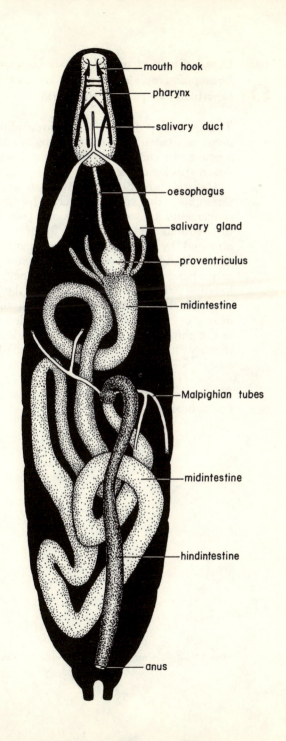

mouth hook

pharynx

salivary duct

oesophagus

salivary gland

proventriculus

midintestine

Malpighian tubes

midintestine

hindintestine

anus

FIGURE

9

Detailed map of one of the six chromosomes of *Drosophila repleta*, a species of fruit fly found in virtually all parts of the world.

This is a copy of one of the finest chromosome maps in existence; maps showing a comparable amount of detail are available for fewer than a dozen species of flies.

The structural details shown here are constant throughout the species. Except for gross rearrangements known as inversions (see Fig. 18), this chromosome of a larval fly of this species, obtained by mating two adults captured anywhere in the world, will appear as shown in this map, band for band. The constancy of these banding patterns is assured only by the precision with which chromosomes duplicate themselves during cell division. Identically banded segments in the chromosomes of two individuals owe their identity to descent from an ultimate, single source.

(Because of limited space, the chromosome has been drawn in short segments.)

After Wharton; courtesy of the University of Texas.

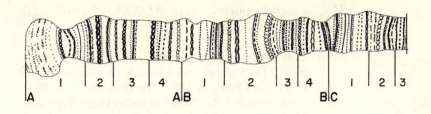

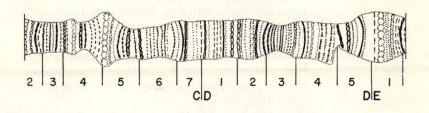

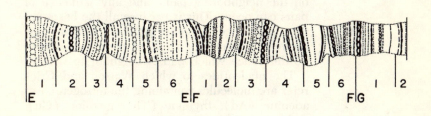

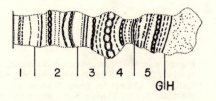

FIGURE

10

A schematic representation of the structure of deoxyribonucleic acid (DNA), the physical basis of heredity.

This model was proposed by an American biologist, J. D. Watson, and a British physical chemist, F. H. C. Crick, in 1953; for developing it they were awarded the Nobel Prize in Physiology and Medicine in 1962. As elegant an explanation of the model as one can find has been given by the two scientists themselves:

"It should be emphasized that since either base can form hydrogen bonds at a number of points one can pair up *isolated* nucleotides in a large variety of ways. *Specific* pairing of bases can only be obtained by imposing some restriction, and in our case it is in direct consequence of the postulated regularity of the phosphate-sugar back bone.

"It should be further emphasized that whatever pair of bases occurs at one particular point in the DNA structure, no restriction is imposed on the neighboring pairs, and any *sequence* of pairs can occur. This is because all the bases are flat, and since they are stacked roughly one above the other like a pile of pennies, it makes no difference which pair is neighbor to which."

The nucleotides to which Watson and Crick refer are molecules containing the nitrogen bases adenine (Ad), thymine (Th), guanine (Gu), and cytosine (Cy).

Figure adapted from Maurice Sussman, *Growth and Development*, 2nd. © 1964. Reproduced by permission of Prentice-Hall, Inc. Quotation from Watson and Crick; courtesy of the Cold Spring Harbor Symposium on Quantitative Biology.

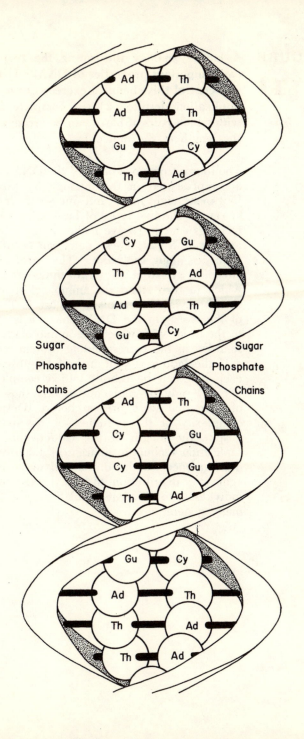

Sugar

Phosphate

Chains

Sugar

Phosphate

Chains

FIGURE

11

**A ladder analogy illustrating the two geneti-
cally important properties of DNA: (1) the precise control this chemical exercises over its own
duplication despite (2) the complete freedom
with which pairs of bases can be arranged along
the length of the molecule.**

The four nitrogenous bases of the DNA molecule
are represented in the analogy by three-inch,
four-inch, eight-inch, and nine-inch fragments
of rungs. The two types of base-pairs that result
in rungs exactly twelve inches long are 3-plus-9
and 4-plus-8 (a). These pairs can occur in any
sequence along the ladder. They are used as a
chemical alphabet to write three-letter words
which in turn stand for amino acids, much as
the abbreviated symbols of Figure 2. The order
of the words in the DNA molecule is identical
to the order with which the amino acids are assembled in the corresponding protein molecule.

If the ladder is split in half lengthwise (b),
each half can be rebuilt into a complete ladder (c) only by replacing exactly those pieces
of rung that were present before. Despite the
complexity of the sequence of base-pairs and
despite the length of the ladder, this precise
replication technique guarantees that the original sequence can be duplicated over and over
again for distribution to many cells, to many
individuals, to different populations, and through
many generations.

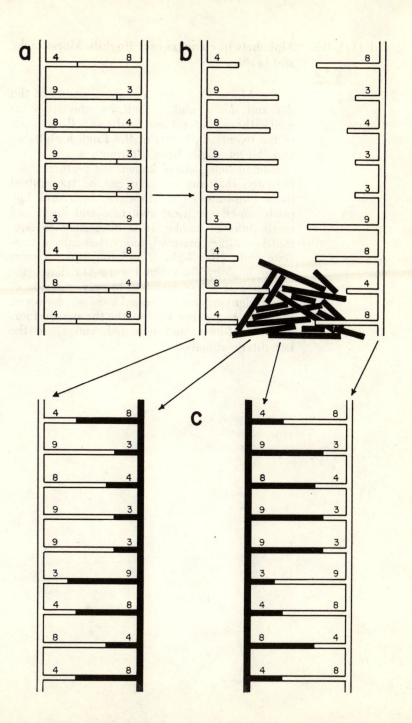

FIGURE

12

Alphabets in common use: English, Morse code, and braille.

In the Morse Code, the alphabet consists of the *dot* and *dash* (and, technically, the *space* as well); these "letters" are used to spell out each of the twenty-six letters of the English alphabet and the ten digits from 0 through 9.

Human beings have known for centuries, apparently, that any message can be transmitted by a two-state code: dot-dash, high and low pitch, on-off electrical switches, and large and small puffs of smoke. It is interesting to note that hereditary material, too, is basically a two-state code (the 3-plus-9 and the 4-plus-8 pairs in the analogy shown in Figure 11). Since the sequence of pieces attached to only one side of the ladder contains the coded message, however, there are really four letters (the three-inch, four-inch, eight-inch, and nine-inch rungs) in the hereditary alphabet.

			MORSE CODE	BRAILLE
A	a	*a*	· —	
B	b	*b*	— · · ·	
C	c	*c*	— · — ·	
D	d	*d*	— · ·	
E	e	*e*	·	
F	f	*f*	· · — ·	
G	g	*g*	— — ·	
H	h	*h*	· · · ·	
I	i	*i*	· ·	
J	j	*j*	· — — —	
K	k	*k*	— · —	
L	l	*l*	· — · ·	
M	m	*m*	— —	
N	n	*n*	— ·	
O	o	*o*	— — —	
P	p	*p*	· — — ·	
Q	q	*q*	— — · —	
R	r	*r*	· — ·	
S	s	*s*	· · ·	
T	t	*t*	—	
U	u	*u*	· · —	
V	v	*v*	· · · —	
W	w	*w*	· — —	
X	x	*x*	— · · —	
Y	y	*y*	— · — —	
Z	z	*z*	— — · ·	

FIGURE

13

Semidiagrammatic representation of human chromosomes.

In each grouping the members of each chromosome pair are drawn side by side, and all pairs are arranged in sequence according to size. (The individual members of each pair are themselves doubled as a result of the experimental technique used in making the chromosomal preparations.) The chromosomes represented in (a) are those of a normal male; in (b), those of a normal female. The sexes differ only in the presence of an X and a Y chromosome in males and of two X chromosomes in females. The chromosomes of a male mongoloid idiot are represented in (c); this set of chromosomes differs from that of a normal male (a) by the presence of *three* rather than *two* chromosomes numbered 21.

The differences between men and women illustrate the enormous effect a single chromosome can exert on normal development. The abnormalities, mental and physical, exhibited by mongoloid idiots illustrate the severe pathological effects wrought by a seemingly minor disturbance in the normal chromosomal endowment of an individual.

(a, b): After Victor A. McKusick, *Human Genetics,* © 1964. Reproduced by permission of Prentice-Hall, Inc. (c): After McKusick, *Medical Genetics 1958–1960,* St. Louis, 1961, The C.V. Mosby Company.

a

1 2 3 4 5

6 7 8 9 10 11 12 13 14 15

16 17 18 19 20 21 22 X Y

b

1 2 3 4 5

6 7 8 9 10 11 12 13 14 15

16 17 18 19 20 21 22 X

c

1 2 3 4 5

6 7 8 9 10 11 12 13 14 15

16 17 18 19 20 21 22 X Y

FIGURE

14

Giant chromosomes from two different larval tissues of the tropical fungus fly, *Rhynchosciara angelae.*

The upper drawing in each of the three sections is of one particular chromosome (chromosome A) as it appears in salivary-gland cells; the lower drawing is of the same chromosome as it appears in the cells of the Malpighian tubules. (See Figure 8 for a diagram of the anatomy of a larval fly.) The symbols used in these drawings identify corresponding landmarks that are visible in the chromosomes of the two tissues.

The precise replication of chromosomal material during cell division is demonstrated not only by identical banding patterns of different cells of the same tissue but also by the nearly identical patterns in cells of widely separated tissues as well. Indeed, the precise pairing of chromosomes obtained by a larva from two different, and most likely unrelated, individuals— father and mother—illustrates the great precision with which chromosomes are duplicated.

(Limited space has made it necessary to draw these chromosomes in three segments.)

After Pavan and Breuer; courtesy of the *Journal of Heredity.*

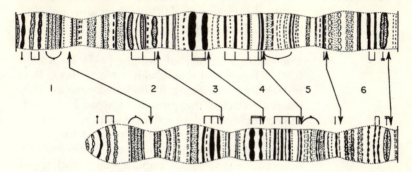

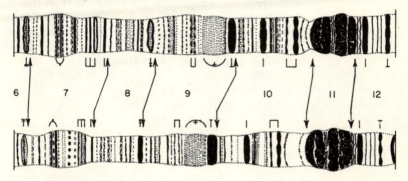

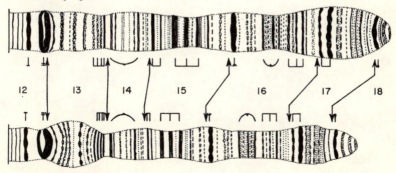

FIGURE

15

Variation in the appearance of one region of a giant chromosome of *Rhynchosciara angelae*, the Brazilian fungus fly.

The entire sequence of puffing (a–j) illustrated in this figure requires about sixteen days. It is revealed by sacrificing individual larvae at daily intervals during the growth of what is a naturally synchronized brood of larval maggots.

Some chromosomal puffs represent times and regions of intense gene activity. Since puffs in different chromosomal regions come and go at different times in a given tissue, and since the same chromosomal region puffs at different times in different tissues, it appears that the genes actively initiating protein synthesis differ in time and in location within an individual. Identical chromosomal complements in all cells of the body do not necessarily mean, therefore, identical gene activities in all of these cells.

After Breuer and Pavan; courtesy of Springer-Verlag.

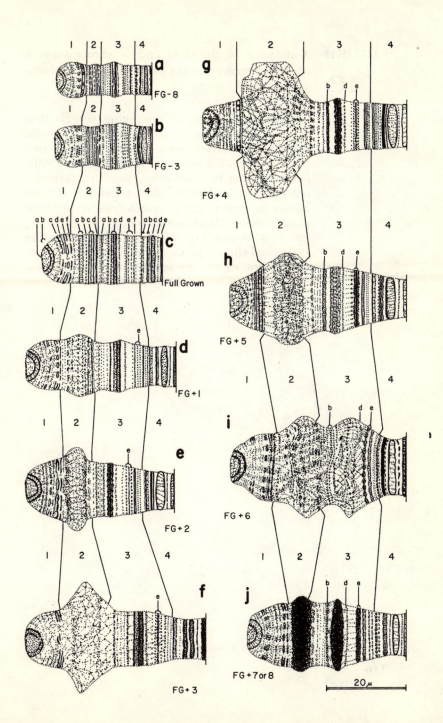

20μ

FIGURE

16

A drawing of the right end of chromosome 2 of the common vinegar fly, *Drosophila melanogaster*.

Careful examination of this figure reveals that what appears to be a single structure is really double. The chromosomes derived from the two parents are coiled loosely around each other with corresponding bands matched perfectly throughout their entire length. Most geneticists do not bother to represent the double nature of these giant salivary-gland chromosomes (for example, see Figs. 9 and 14). The ease with which the individual chromosomes can be seen varies somewhat from species to species.

After Bridges; courtesy of the *Journal of Heredity*.

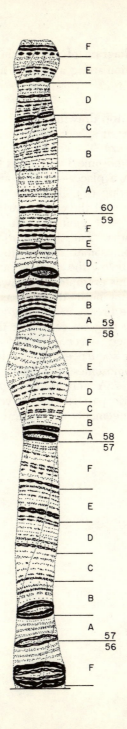

FIGURE

17

A diagram of the origin of a chromosomal inversion using a map of a giant chromosome for clarity.

The top drawing (a) shows chromosome 3 of a western species of fruit fly, *Drosophila pseudo-obscura*. The center drawing (b) is the identical chromosome 3, but it has been drawn with breaks in two places—at 70B and at 76B. The bottom drawing (c) illustrates the banding pattern one obtains if the center portion, 70B–76B, is rotated 180° and rejoined to the two terminal segments. (Notice that the numbers which identify the chromosomal regions in the inverted segment are below the drawing and in reverse order.)

This example is somewhat artificial. Inversions do not normally arise as the result of chromosome breaks in salivary-gland cells; instead they arise in germ cells. The illustration shows that a trained observer could recognize the altered gene sequence by a careful examination of giant chromosomes; the heavy bands in regions 72A and 73A and B, for example, form an asymmetric pattern that serves to contrast the gene arrangements in the top and bottom drawings.

FIGURE

18

A schematic representation of the origin of inversions (compare with Figure 17) and the pairing of chromosomes that differ by an inverted segment.

The four inversions shown here are all simple two-break inversions; they differ only in their positions on the chromosome.

The formation of a loop during the pairing of two chromosomes that differ by an inverted segment is an inescapable geometric outcome of the attraction of corresponding bands of the two chromosomes for one another, despite their different locations along the lengths of the chromosomes.

After Dobzhansky; courtesy of the Carnegie Institution of Washington.

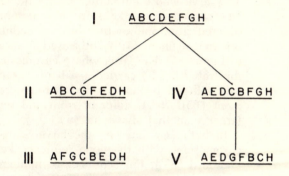

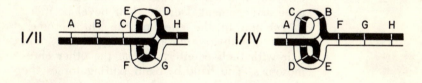

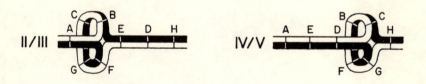

FIGURE

FIGURE

19

Drawings of inversion loops in the giant chromosomes of the fruit fly, _Drosophila pseudoobscura._

The larva whose chromosomes are illustrated in (a) carried a pair of third chromosomes that differed from one another by the medium-sized inverted segment (70B–76B) used as an example in Figure 17. The larva whose chromosomes are illustrated in (b) carried a pair of chromosomes that differed by an inverted chromosomal segment (65B–76A), different from and somewhat larger than that shown in (a).

In both (a) and (b) the larvae carried one chromosome with the gene sequence known as Standard (ST); this sequence is numbered consecutively from 63 through 81. The two inverted sequences in (a) and (b) are known as Arrowhead (AR) and Pikes Peak (PP). Drawing (c) shows the pairing pattern observed in a larva carrying one Pikes Peak and one Arrowhead gene arrangement. Despite their novel positions in relation to the ends of the chromosomes, bands of salivary chromosomes seek out and pair with their counterparts on the other chromosomes. The band-by-band pairing forces the chromosomes into complex loop formations—the more dissimilar the sequences, the more complex the pairing configuration.

After Dobzhansky; courtesy of the Carnegie Institution of Washington.

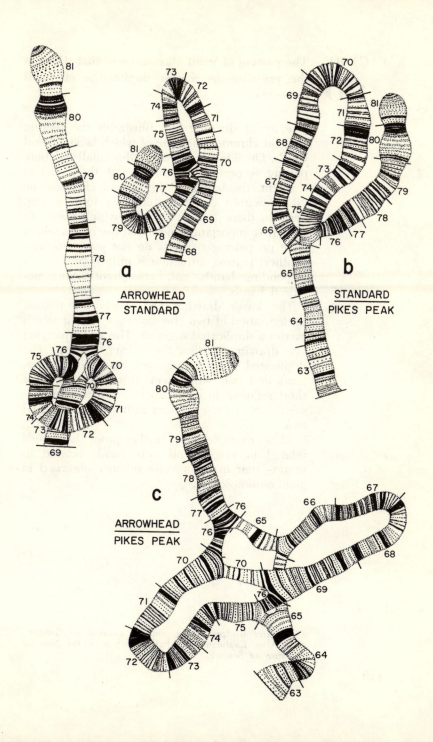

ARROWHEAD
STANDARD

a

STANDARD
PIKES PEAK

b

ARROWHEAD
PIKES PEAK

c

FIGURE

20

The pairing of giant chromosomes that differ in the presence-absence or duplication of small segments.

The upper drawing (a) illustrates the pairing of giant chromosomes one of which lacks several bands. Outside the region of this small deletion, pairing is perfect; at the site of the deficient segment, the bands of the normal chromosome (those within the indicated arc) form a loop because there is nothing with which they can pair. It is important to note that the bands which have no pairing partners do not pair with one another; pairing attraction is only between corresponding bands, *not* between otherwise unpaired bands.

The lower drawing (b) shows the pairing configuration of two chromosomes, one of which carries a duplicated segment. The accompanying line drawing clarifies the pairing pattern. The duplicated bands simply seek out their counterparts and pair with them; this gives rise to a short segment in which there are three chromosome parts joined together rather than the usual two.

These examples offer further proof that bands related in origin—and only bands related in origin—pair in the precise manner observed in giant chromosomes.

(a): After Bridges, Skoog, and Li; courtesy of *Genetics*.
(b): After Kaufmann and Bate; courtesy of the National Academy of Science.

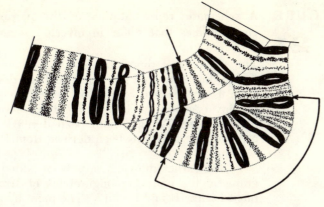

a

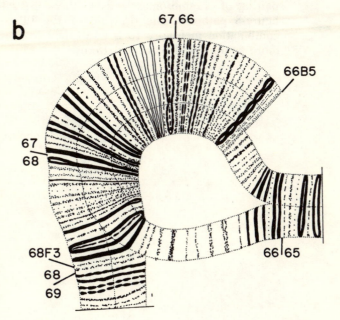

b

67,66

66B5

67
68

68F3
68
69

66 65

FIGURE

21

A schematic representation of the pairing of chromosomes that differ in inverted segments.

The top part of this figure is identical to Figure 18. Each inverted sequence has been derived from its predecessor by two breaks, followed by the inversion of the interstitial piece. Arrangements II and IV have both been derived from arrangement I, III has been derived from II and V from IV.

To the earlier drawing have been added diagrams of the pairing configurations of arrangement I with III and I with V. The point by point pairing of chromosomes leads in these cases to figure-8 patterns. The drawing of the PP/AR combination in Figure 19 is equivalent to these more complicated pairing patterns.

After Dobzhansky; courtesy of the Carnegie Institution of Washington.

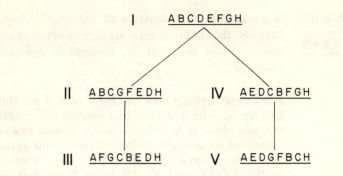

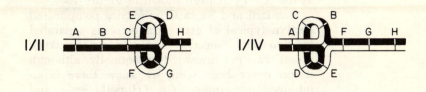

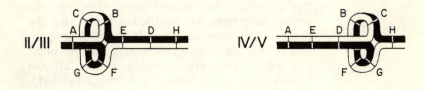

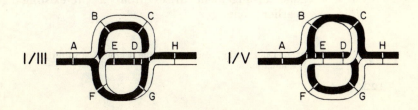

FIGURE

22

A composite reconstruction of the relationships between the different gene arrangements of chromosome 3 of the fruit fly, *Drosophila pseudo-obscura*.

Each gene arrangement has been named for the locality in which it was first discovered (with the exception of Standard). Gene arrangements separated by a single double-headed arrow form simple inversion loops with one another (similar to the ST/AR and ST/PP loops illustrated in Figure 19) in salivary-gland cells. Arrangements separated by two double-headed arrows give figure-8 pairing patterns with one another (such as the AR/PP pattern shown in Figure 19).

Standard and Santa Cruz form a complicated pattern typical of gene arrangements separated by two double-headed arrows (four breakage points, two per arrow). Consequently, although it has never been seen, we know there is an intermediate arrangement (Hypothetical) and we know precisely how its bands are arranged.

Although all the arrows have been drawn with two heads, something *is* known about the direction in which they should point. However, the arrows have been left double-headed so that the reader might convince himself that the designation of any one arrangement as a starting point (primitive gene arrangement) causes all the arrows to become unidirectional (single-headed). The theoretical importance of these gene arrangements lies in the sequences they form, not in the particular direction the arrows point. The sequences are, of course, sequences in time as well as in geometric patterns; any one gene arrangement arises from a pre-existing (earlier) one.

After Dobzhansky; courtesy of the Carnegie Institution of Washington.

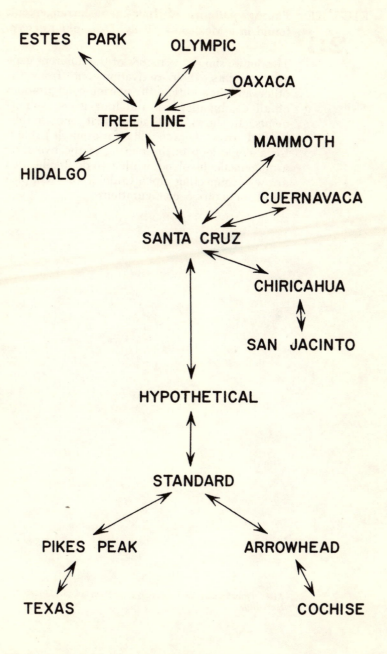

FIGURE
23

Pairing patterns of four gene arrangements found in a Mexican fruit fly, *Drosophila azteca*.

This figure shows the names of the different gene arrangements, their derivation one from the other, and, in addition, the pairing configurations of all combinations of the four gene arrangements. It shows once more that one double-headed arrow (beta to delta, for example) stands for a single loop in the pairing of the two gene arrangements involved, while two double-headed arrows (connecting alpha to delta, for example) stand for figure-8 configurations.

After Dobzhansky and Sokolov; courtesy of the *Journal of Heredity*.

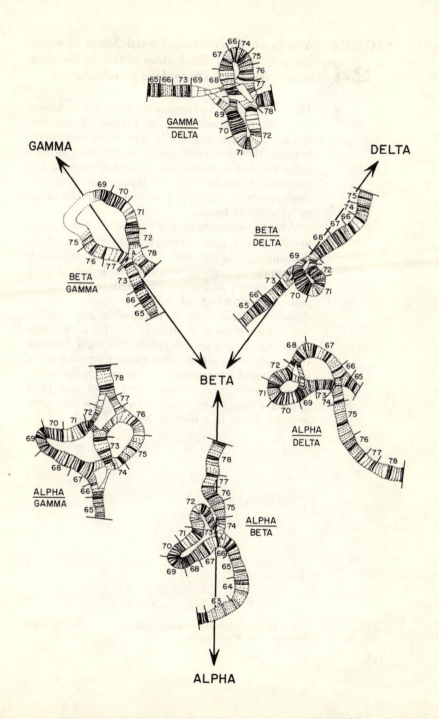

FIGURE

24

A map of the western United States showing the geographical distribution of four of the gene arrangements of *Drosophila pseudoobscura.*

These four gene arrangements have been chosen from among those listed in Figure 22 not because they are especially common, nor even because they are particularly widespread in distribution, but because the areas they occupy are disjunct. Actually, each of these four gene arrangements is relatively rare in many of the areas in which it is found.

When geneticists first observed that the same non-Standard gene arrangement was found in flies living throughout an area encompassing thousands of square miles, they believed that the repeated, independent origin of the same gene arrangement at different times and in different places was a reasonable explanation for the observed distribution. As knowledge about inversions accumulated, this view became untenable. It is much more likely that each gene arrangement has arisen *once,* in *one* chromosome of *one* individual at *one* time, and has then spread through the species to occupy the geographic areas in which it is now found.

After Dobzhansky and Epling; courtesy of the Carnegie Institution of Washington.

OLYMPIC   – – – – –
TREE LINE   ·············
ESTES PARK   ·–·–·–·
SANTA CRUZ   ———

FIGURE

25

A checkerboard showing the difficulties that would be encountered by a population of organisms in which different individuals had different numbers of chromosome pairs.

The association of chromosomes in pairs is usually a necessary step in successful germ-cell formation. Their orderly separation during cell division is assured by this association. Consequently, the bulk of all progeny arising in the "checkerboard" population shown here would be at least partially sterile. Only the offspring marked with an asterisk would be fully fertile like the original parents. Unequal initial frequencies of the four types of chromosome numbers (one, two, three, and four pairs) and the infertility of most individuals arising through random mating would lead eventually to the establishment of one chromosomal number as the *only* number in the population. This, of course, is what one normally finds.

♂ MALE FEMALE ♀	A B C D / A B C D	A B C D / A B C D	A B C D / A B C D	A B C D / A B C D *
A B C D / A B C D	A B C D / A B C D	A B C D / A B C D	A B / A B C D / C D	A B / A B C D / C D
A B / A B C D / C D	A B C D / A B C D *	A B C D / A B C D	A B / A B C D / C D	A B / A B C D / C D
A B / A B C D / C D	A B / A B C D / C D	A B C D / A B C D *	A B / A B C D / C D *	A B / A B C D / C D
A B C D / A B C D	A B C D / A B C D	A B C D / A B C D	A B C D / A B C D	A B C D / A B C D

FIGURE

26

The use of overlapping deficiencies to determine the physical location of genes relative to individual bands of giant salivary-gland chromosomes.

The chromosomal segment shown here is part of the X chromosome of the vinegar fly, *Drosophila melanogaster*. The numerical symbols in the left-hand column designate those chromosomes, obtained experimentally, from which certain bands are missing (as in the case illustrated in Figure 20); the black bar to the right of each numerical symbol indicates the missing bands.

Deletions 258–11 through 264–31 are known from genetic tests (see p. 37) to include the gene for white eyes. Consequently, this gene is on, or adjacent to, band 3C1, the only band common to these seven deficiencies. The eleven deficient chromosomes, from N–8 through 264–19, include a second gene known as "facet" (mutations at this locus disrupt the structure of the eye). This gene must be on or near band 3C7, a band that is missing from all eleven of these deficient chromosomes.

Once in a while a chromosome seems to have a deletion on the basis of a genetic test, but microscopic examination fails to reveal it; this need not disturb us unduly because (1) human eyes are not infallible, (2) microscopes, even the best ones, have optical limitations, and (3) invisible gene mutations caused by alterations at the molecular level may produce effects indistinguishable from those of deficiencies.

After Slizynska; courtesy of *Genetics*.

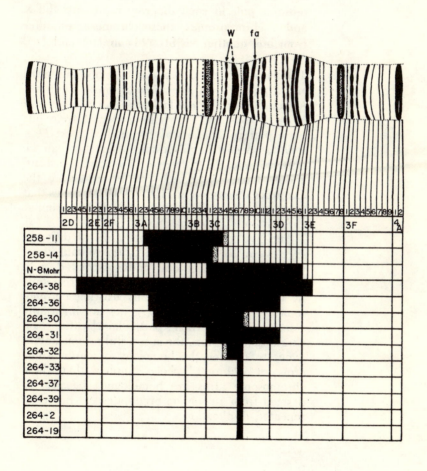

Diagrams of the chromosomal complements of several species of *Drosophila*: (a) *D. virilis,* (b) *D. pseudoobscura,* (c) *D. melanogaster,* and (d) *D. willistoni.* The bottommost chromosome pair in each diagram represents the X and Y chromosomes; these chromosomes differ from one another visibly only in (b) and (c).

The basic pattern of five long chromosomal arms and a pair of dots is obvious in these diagrams. Although *D. willistoni* lacks the pair of dots, and other species, not shown here, depart from this basic pattern in one way or another, most *Drosophila* species have chromosomes that can be resolved into the pattern of five long arms and a dot. Thus, *Drosophila* can be added to the examples illustrated in Figure 6 which shows that taxonomically similar species tend to have similar chromosome complements.

The five long chromosomal arms of different species can be homologized by the mutant genes with similar effects one finds associated with them. Within species, chromosomes arise only by the duplication of already existing chromosomes. Within species, too, genes are invariably associated with their particular locations on chromosomes. Thus, the successful use of mutant genes to deduce the homology of chromosome arms of different species implies that these chromosome arms are homologous because of descent from a common origin, descent based on repeated duplication.

Adapted from *Evolution in the Genus Drosophila* by Patterson and Stone with permission of The Macmillan Company. Copyright by The Macmillan Company.

a *D. virilis*

b *D. pseudoobscura*

c *D. melanogaster*

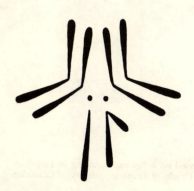

d *D. willistoni*

FIGURE

28

An illustration of "instant" speciation based on the radish, the cabbage, and their hybrid.

Both the radish (a) and the cabbage (d) have eighteen chromosomes in cells other than germ cells. These eighteen chromosomes represent nine pairs; during germ-cell formation the halving of the chromosome number proceeds smoothly in each species. High fertility is reflected in the large seed pods.

The hybrid (b) produced by crossing the radish and the cabbage also has eighteen chromosomes; each parental germ cell contributes nine. These chromosomes are not pairs, and, consequently, they do not separate regularly during germ-cell formation. The resulting imbalance in chromosome number causes a failure in the seed set of these plants; hence the seed pod is small.

If a hybrid plant is made to undergo chromosome doubling by an appropriate experimental treatment, the resulting plant (c) has thirty-six chromosomes, eighteen radish and eighteen cabbage. Germ-cell formation now proceeds regularly since each chromosome is a member of a pair, and one member of each pair goes regularly to each daughter cell during the reduction division.

After Karpechenko, based on a figure appearing in Dobzhansky, *Genetics and the Origin of Species;* courtesy of Columbia University Press.

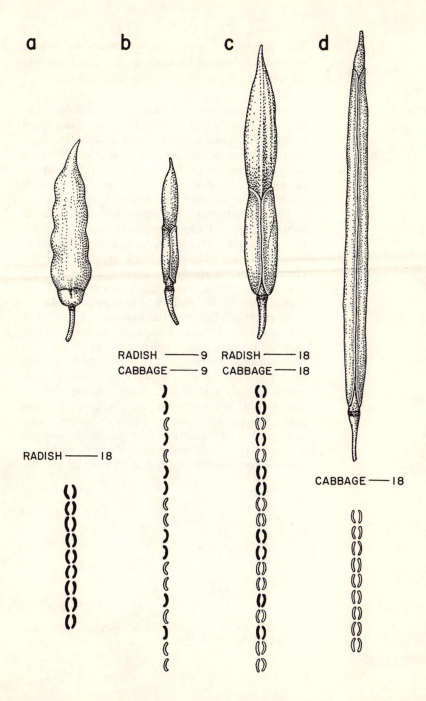

a

b

c

d

RADISH —— 9
CABBAGE —— 9

RADISH —— 18
CABBAGE —— 18

RADISH —— 18

CABBAGE —— 18

FIGURE

29

The pairing of giant chromosomes in hybrid individuals formed by mating different *Drosophila* species.

The chromosomes of *melanogaster-simulans* hybrids (a) pair almost perfectly throughout their length. The chromosome illustrated is the right arm of chromosome 3. The two species differ by a large inversion in this chromosome and by an occasional minor difference in banding pattern such as that indicated by arrows.

The chromosomes of *pseudoobscura-miranda* hybrids (b) pair much less perfectly than those of the *melanogaster-simulans* hybrids. Nevertheless, there are relatively large segments in the chromosomes of these two species that match up and pair tightly; some of these segments are sections 88–89, 96, and 99 of chromosome 4.

In previous figures (for example, Fig. 20) it was shown that only bands of common origin pair in salivary-gland cells of *Drosophila*. The chromosomal pairing shown here, then, implies that different species share chromosomal segments of common origin. Such sharing can be true only if what was once a single species has given rise to two or more of today's species. This is one type of evolution: the origin of species. Another type of evolution, change within a species, has been illustrated in Figures 22 and 23.

(a): After Patau. (b): After Dobzhansky and Tan. Both courtesy of Springer-Verlag.

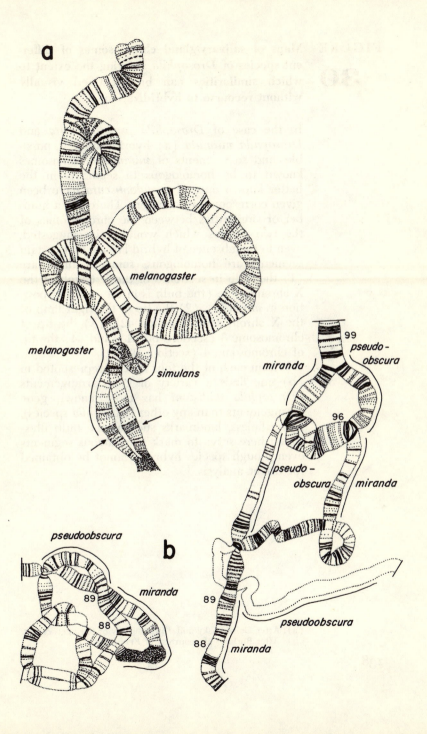

a

melanogaster

melanogaster

simulans

99 pseudo-
obscura

miranda

96

pseudo-
obscura   miranda

b

pseudoobscura

miranda

89

88

89
miranda

88
miranda

pseudoobscura

FIGURE

30

Maps of salivary-gland chromosomes of different species of *Drosophila* showing the extent to
which similarities can be identified visually
without recourse to hybridization.

In the case of *Drosophila pseudoobscura* and
Drosophila miranda (a) hybridization is possible, and so segments of *miranda* chromosomes
known to be homologous to segments in the
better known ones of *pseudoobscura* have been
given corresponding numbers. There are a number of similarities between the chromosomes of
the two species which would have suggested,
even in the absence of hybrid larvae, that certain
segments are homologous; some of these are
(1) the bulb in section 9 of the left limb of the
X chromosome (the bulb is in the inverted position in *miranda*), (2) the tip of the right limb of
the X chromosome (section 42), (3) the tip of
chromosome 3 (sections 80–81), and (4) the tip
of chromosome 4 (sections 98–99).

Within each of the four species represented in
(b) one finds a variety of gene arrangements
(*Drosophila willistoni* has more known gene
arrangements than any other *Drosophila* species).
Nevertheless, landmarks such as the bulb illustrated here serve to mark homologous segments
even though species hybrids cannot be obtained
for direct analysis.

(a): After Dobzhansky and Tan; courtesy of Springer-Verlag.
(b): After Burla *et al.*; courtesy of *Evolution.*

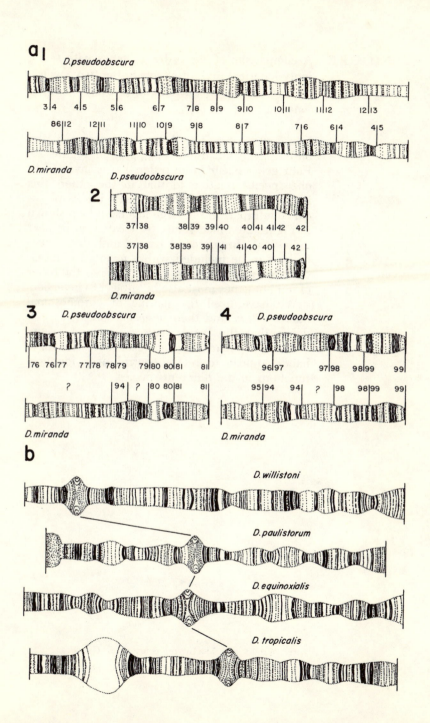

FIGURE

31

A comparison of the order of genes with seemingly identical mutant effects in the two species, *Drosophila melanogaster* and *Drosophila simulans*.

These genes are found on chromosome 3, the chromosome illustrated (in part) in Figure 29. Four genes occur in an order that is inverted in one species relative to that in the other; this inverted sequence of mutant genes has its physical counterpart in the large inversion loop shown in Figure 29. Since genes are located on or near individual chromosomal bands, and since these bands can be included within inverted chromosomal segments, then—as we see here—the order of mutant genes must also be subject to inversion.

We have used the pairing of chromosomal bands *within* and then *between* species as evidence that the bands have a common origin. We now see that the genes associated with these bands appear to retain control over identical functions in the different species.

After Wright and MacIntyre; courtesy of *Genetics*.

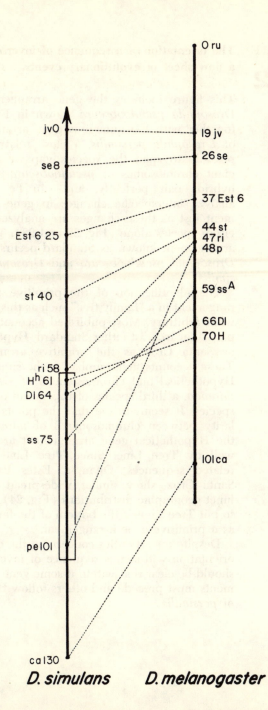

D. simulans	D. melanogaster
	0 ru
jv 0	19 jv
se 8	26 se
	37 Est 6
Est 6 25	44 st
	47 ri
	48 p
st 40	59 ss^A
	66 Dl
	70 H
ri 58	
H^h 61	
Dl 64	
ss 75	
	101 ca
pe 101	
ca 130	

FIGURE

32

The orientation of a sequence of inversions into a flow sheet of evolutionary events.

This figure includes the gene arrangements of *Drosophila pseudoobscura* shown in Figure 22 together with the numerous gene arrangements of *Drosophila persimilis,* a close relative. It is possible to span the species gap because the giant chromosomes of *pseudoobscura-persimilis* hybrids pair perfectly, and can be used to analyze interspecific changes in gene arrangement just as these changes are analyzed within either species alone (Fig. 22 and 23). The gene arrangement known as Standard occurs in both *Drosophila pseudoobscura* and *Drosophila persimilis.*

The identification of the primitive gene arrangement in a "family tree" such as this is somewhat arbitrary. Most published accounts of this particular tree list either Standard, Hypothetical, or Santa Cruz as the primitive arrangement. These accounts, too, emphasize the similarity of Hypothetical and chromosome 3 of *Drosophila miranda,* a third species of this group of related species. It seems to me that the points of similarity between chromosome 3 of *miranda* and the Hypothetical gene arrangement are met as well by Tree Line. Since Tree Line and its related sequences, Olympic, Estes Park, and Santa Cruz, show similar widespread but disjunct geographic distributions (Fig. 24), I prefer to put Tree Line at the bottom of the family tree as a primitive gene arrangement.

Despite uncertainties concerning the details of orientation within this sequence of inversions, it should be clear that within it some gene arrangements must precede and others follow the origin of *persimilis.*

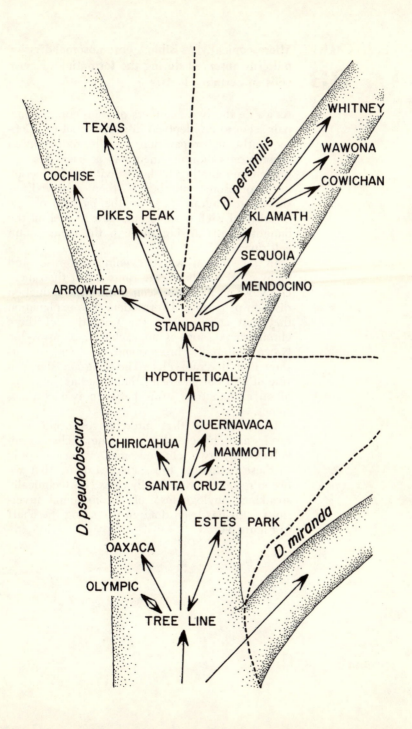

FIGURE

33

Microscopically visible chromosomal-division patterns observed during the formation of germ cells in certain plants.

As a rule the two members of each chromosome pair seem to be identical in gross structure. Consequently, in preparation for the reduction in chromosome number, members of each pair approach one another without dividing. Reduction in chromosome number is then achieved, as shown in diagram (a), by the passage of one member of each chromosome pair to each of the daughter cells arising through the concomitant cell division.

In the Jimson weed, evening primrose, and many other plants, zigzag rings rather than mere chromosome pairs are seen during the reduction division of germ-cell formation. The explanation that best accounts for the formation of these chromosomal rings is an exchange of arms between different chromosomes of one set, as shown in diagram (b). The diagram shows a ring of four chromosomes that arises as the result of an exchange of arms between two chromosomes.

If we assume that interchanges occur very rarely and, hence, that any one exchange involves only two chromosomes, we can apply to these exchanges the same reasoning that we use in reconstructing family trees (or, technically speaking, phylogenies) of chromosomal inversions. This has been done in preparing the chart in Figure 34.

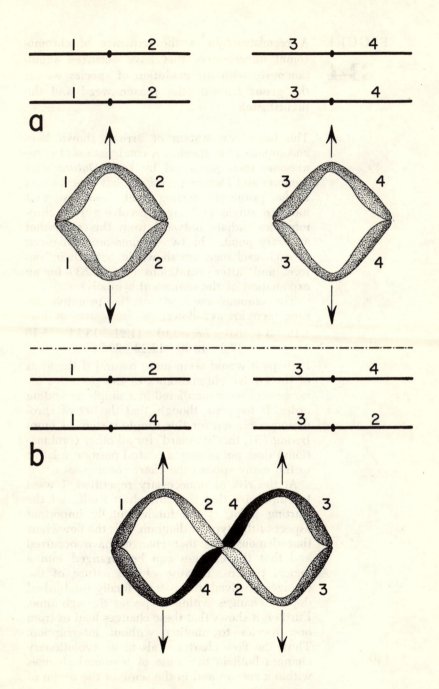

FIGURE

34

A reconstruction of the sequence of chromosomal interchanges that have occurred simultaneously with the evolution of species within the genus *Datura* (the Jimson weed and the nightshade).

The branching system of arrows shown here encompasses ten species. A combination of chromosome ends possessed by both *Datura stramonium* and *Datura quercifolia* has been chosen as the "primitive" starting point; the arrows that indicate single exchanges involving two chromosomes radiate outward from this somewhat arbitrary point. The two chromosomes involved in each exchange are shown in both their "before" and "after" conditions (see p. 53) for an explanation of the numerical symbols).

The chromosome ends of the primitive arrangement are as follows:

1.18   3.4   5.6   7.8   9.10   11.21   13.14   15.16
17.2   19.20   12.22   23.24

Perhaps it would seem more natural if the arms of the twelve chromosomes of the primitive arrangement were numbered in a simple ascending order. It happens, though, that the set of chromosomes chosen for this simple system of numbering (S1, the "standard" for all other combinations) does not occupy a central position relative to the many species the chart encompasses.

At the risk of unnecessary repetition, I want to emphasize that the flow chart itself, not the starting point, is the fundamentally important aspect of this type of diagram. It is the flow chart that demonstrates that changes have occurred and that the changes can be arranged into a logical pattern. Because of the nature of the changes involved, the pattern finally established depicts changes within a species through time. Further, it shows that these changes lead us from one species to another without interruption. Thus, the flow chart reveals to us evolutionary changes both in the sense of temporal changes within a species and in the sense of the origin of new species.

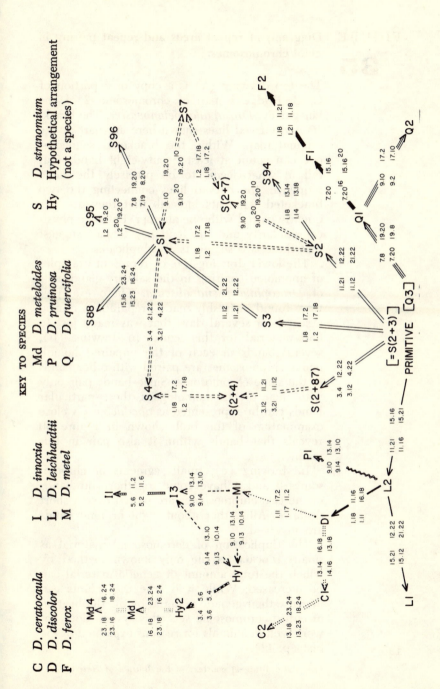

KEY TO SPECIES

C   *D. ceratocaula*
D   *D. discolor*
F   *D. ferox*

I   *D. innoxia*
L   *D. leichhardtii*
M   *D. metel*

Md   *D. meteloides*
P   *D. pruinosa*
Q   *D. quercifolia*

S   *D. stramonium*
Hy   Hypothetical arrangement
     (not a species)

FIGURE

35

Diagrams of repeat areas and repeat pairing in giant chromosomes.

The top drawing (a) is a copy of a portion of C. B. Bridges's map of chromosome 2 of the vinegar fly, *Drosophila melanogaster*. The brackets and curved lines shown here are part of the original map. Within the bracketed segments, one can count at least twenty-eight bands that fall in what seems to be precisely the same sequence; the curved lines connecting the two bracketed segments indicate that they are frequently paired with one another. Both the physical similarity and the tendency to pair suggest that these two segments are duplicates.

The lower drawings (b, c) represent examples of anomolous pairing in the salivary-gland cells of *Drosophila pseudoobscura*. These examples were found in freshly made preparations during a search of several days that was made just to find material for this figure. In drawing (b), several bands in each of the unpaired homologous chromosomes are paired with other bands in the same chromosome. Since bands pair only with their precise homologues, these particular bands can be represented as *abcddcba*. (A close examination of the bulb shown in Figure 30 reveals that bands within it also pair in this manner.)

In drawing (c), small segments of chromosomes 3 and 4 have come together and bands of the two chromosomes are paired over a short distance. Altogether about eight or nine bands are involved.

The duplication of chromosomal material as small "repeats" is the only known method by which the total amount of genetic material can be increased. While it is true that plants can increase their genetic material by the duplication of entire chromosome sets, this system does not work well in animals for reasons explained in the text (p. 42).

(a): After Bridges; courtesy of the *Journal of Heredity*.

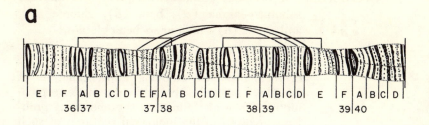

a

E | F | A | B | C | D | E | F | A | B | C | D | E | F | A | B | C | D | E | F | A | B | C | D

36|37   37|38   38|39   39|40

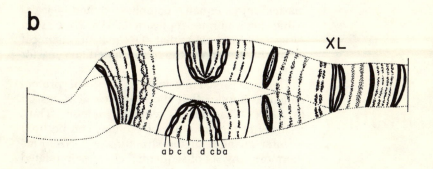

b

XL

a b c d d c b a

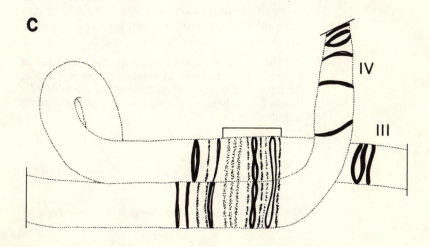

c

IV

III

FIGURE

36

The complete amino-acid sequences of the four subunits—alpha (α), beta (β), gamma (γ), and delta (δ)—of human hemoglobins.

These chains of amino acids illustrate the success chemists are enjoying in the analysis of complex molecules by utilizing modern techniques. They also illustrate the enormous problems involved in the biological synthesis of such molecules. Each molecule is made according to a *precise* formula, not to some vague or ill-defined plan. Consequently, the synthesis of one of these molecules requires the existence of a master copy, equally complicated and yet capable of accurate replication. We have seen (Figs. 10 and 11) that DNA is a chemical admirably suited for the role of hereditary master copy.

The precise composition of a given protein molecule implies that the sequence of rungs on the DNA "ladder" is equally precise. Thus, changes in the structure of proteins can be used to infer changes in the structure of DNA. Furthermore, any set of sequential changes in related protein molecules implies a corresponding set of sequential changes in hereditary material. The latter are evolutionary changes; consequently, a study of protein structure can be used to reconstruct evolutionary changes.

(The heavy lines drawn within the diagrams of the molecules are there to achieve proper spacing in the figure. They have no counterpart in the molecules themselves, where each amino acid is adjacent to its neighbors.)

After Ingram; courtesy of the Columbia University Press.

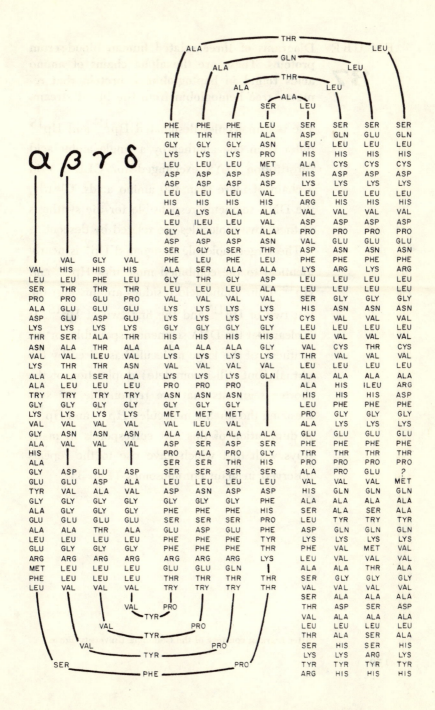

FIGURE

37

Diagrams of three related human blood-serum proteins. These are the alpha chains of amino acids found in haptoglobin, a protein that removes free hemoglobin from the blood stream.

The two molecules designated Hp^{1F} and Hp^{1S} differ from one another by a single amino-acid substitution (LYS exchanged for GLU·N) in a chain of more than 100 amino acids. Clearly, the DNA segments responsible for the synthesis of these two molecules are related by descent.

The larger molecule designated Hp^2 is nearly identical to the combined molecules Hp^{1F} and Hp^{1S}. Four amino-acid residues are missing: the last two of Hp^{1F} and the first two of Hp^{1S}. It is clear that the DNA segment responsible for the synthesis of the large molecule has arisen by the virtual (but still incomplete) duplication of the shorter segments that are responsible for synthesizing the smaller molecules, Hp^{1F} and Hp^{1S}. A duplication of this sort corresponds in many ways with that which gave rise to the repeats observed in giant chromosomes (Fig. 35).

After Ingram; courtesy of the Columbia University Press.

Hp^{IF}

$(1)\cdot(2)\cdot(3)\cdots\cdots LYS\cdots\cdots(n-2)\cdot(n-1)\cdot(n)$

Hp^{IS}

$(1)\cdot(2)\cdot(3)\cdots\cdots GLU\cdot N\cdots\cdots(n-2)\cdot(n-1)\cdot(n)$

Hp^2

$(1)\cdot(2)\cdot(3)\cdots\cdots LYS\cdots\cdots(n-3)\cdot(n-2)\cdot(3)\cdots\cdots GLU\cdot N\cdots\cdots(n-2)\cdot(n-1)\cdot(n)$

FIGURE

38

An attempt to illustrate diagrammatically the evolution of proteins, together with the origin of species.

Any attempt of this sort is destined to be grossly oversimplified. Nevertheless, some such scheme helps us to understand the occurrence of similar proteins in a given individual and the occurrence of nearly identical proteins in members of different species.

On the axis designated "Protein evolution" three sections have been labeled I, II, and III. These represent recognizably different amino-acid chains such as the alpha, beta, and delta chains of hemoglobin; such chains are made by different genes. Presumably, the evolution of proteins, in the sense of changing from a protein that is known by one name to one known by a different name, begins with an abrupt step: the duplication of genetic material (gene duplication). Immediately following gene duplication the proteins synthesized under the control of the two genes must be identical, or very nearly so. Eventually the proteins whose synthesis is controlled by the two duplicate genes diverge in structure because of mutations in the genes themselves. It is the first abrupt step, rather than the gradual divergence following gene duplication, that the figure represents.

Shorter horizontal segments that leave the protein within one or the other of the three classifications represent alterations that do not call for a new designation. For example, nearly all mammals seem to possess several amino-acid chains that are recognized as alpha or as beta chains of hemoglobin despite slight differences between them; these chains are made by mutant forms of the genes involved.

Line segments leading back and to the right represent the origin of new species. A line of

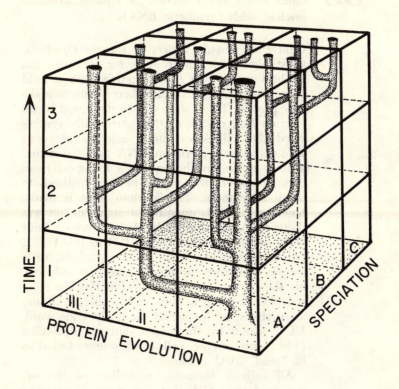

this sort leads from each protein possessed by the parental species at a given moment; that is, at the time of its origin, a new species has virtually the same array of proteins as the species from which it arises. (If it should develop that individual members of a species exhibit substantial protein differences, the statement about the proteins of a newly arisen species would have to be qualified accordingly.)

FIGURE

39

The genetic code as it has been reconstructed from the synthesis of artificial proteinlike molecules under the direction of equally artificial nucleic acids (synthetic RNA).

Cells from a wide variety of sources (bacteria, yeast, algae, rats, and rabbits, for example) can be broken up in a flask together with a mild preservative and still retain their ability to make new protein. This ability can be destroyed by an enzyme that destroys DNA. Artificially synthesized RNA (RNA is a nucleic acid that serves as an intermediary between DNA and the actual protein-making machinery of the living cell) can then be added; this causes protein synthesis to begin once more. The "protein" that is made, however, depends upon the structure of the synthetic RNA. An artificial RNA, for example, built of a simple repetitive plan (...U.U.U.U.U.U..., polyuridylic acid) leads to the synthesis of a simple repetitive "protein" containing only one of the amino acids (...Phe.Phe.Phe.Phe.Phe., polyphenylalanine). Thus, in the absence of any alternatives, we learn that the triplet code word, UUU, stands for the amino acid, phenlyalanine (Phe); furthermore, cells of all sources tested so far "read" UUU as Phe.

All cellular factories, regardless of the organism from which they are derived, interpret the instructions contained in artificial RNAs in the same manner. The genetic code, in other words, appears to be universal. There are literally billions of ways in which nucleic acid code words could have been assigned to the twenty amino acids. That all living organisms use precisely the same code is elegant evidence that these organisms have arisen from but one source, that life on earth has had but a single source from which all present forms have evolved.

After Speyer *et al.*; courtesy of the Cold Spring Harbor Symposium on Quantitative Biology.

Amino Acid   Code Triplets

Amino Acid				
Alanine	CAG	CCG	CUG	
Arginine	GAA	GCC	GUC	
Asparagine	CAA	CUA	UAA	
Aspartic Acid	GCA	GUA		
Cysteine	GUU			
Glutamic Acid	AAG	AUG		
Glutamine	AAC	UAC		
Glycine	GAG	GCG	GUG	
Histidine	ACC	AUC		
Isoleucine	AAU	CAU	UUA	
Leucine	CCU	UAU	UGU	UUC
Lysine	AAA	AUA		
Methionine	AGU			
Phenylalanine	UCU	UUU		
Proline	CAC	CCC	CUC	
Serine	ACG	CUU	UCC	
Threonine	ACA	CCA	UCA	
Tryptophan	UGG			
Tyrosine	ACU	AUU		
Valine	UUG			

References and
Acknowledgments

The material presented in this book is not drawn from an isolated or unique field of research; it represents the backlog of information accumulated by any geneticist interested in evolutionary processes. Nevertheless, the reasonably serious reader should be provided with a list of texts or sources that can be consulted for additional information. To accomplish this task, I have listed below the sources I have consulted or from which I have adapted material for the 39 figures. I do not mean to imply that these are the only possible sources for this illustrative material; they were the sources within arm's length of my desk at the time I was preparing the manuscript. In most cases I have acknowledged the original research papers; in several instances I have referred as well to advanced texts or monographs that give thorough "second-hand" accounts of the original research.

It is a pleasure to acknowledge here the kindness of both authors and publishers in granting permission to use the illustrative material appearing in this book.

Fig. 2. I. H. Herskowitz. 1962. *Genetics.* Boston, Mass.: Little, Brown and Co.

E. H. White. 1964. *Chemical Background for the Biological Sciences.* Englewood Cliffs, N.J.: Prentice-Hall.

Chromosomes, Giant Molecules, and Evolution

Fig. 3. C. B. Anfinsen. 1959. *The Molecular Basis of Evolution.* New York: John Wiley and Sons.

Fig. 4. E. W. Sinnott, L. C. Dunn, and Th. Dobzhansky. 1958. *Principles of Genetics.* New York: McGraw-Hill Book Co.

Fig. 5. *See 4.*

Fig. 6. (a) S. D. McKelvey and K. Sax. 1933. "Taxonomic and cytological relationships of Yucca and Agave." *Journal of the Arnold Arboretum* 14:76–81.

(b) H. E. Warmke. 1937. "Cytology of the Pacific Coast Trilliums." *American Journal of Botany* 24:376–383.

(c) B. Chiarelli. 1962. Comparative morphometric analysis of primate chromosomes. I. The chromosomes of anthropoid apes and man. *Caryologia* 15:99–121.

Fig. 7. J. T. Patterson. 1943. The *Drosophilidae* of the Southwest. *University of Texas Publications* 4313:7–216.

Fig. 8. M. Demerec. 1950. *Biology of Drosophila.* New York: John Wiley and Sons.

Fig. 9. L. T. Wharton. 1942. Analysis of the *repleta* group of *Drosophila. University of Texas Publications* 4228:23–52.

Fig. 10. M. Sussman. 1964. *Growth and Development,* 2d ed. Englewood Cliffs, N.J.: Prentice-Hall.

J. D. Watson and F. H. C. Crick. 1953. The structure of DNA. *Cold Spring Harbor Symposium on Quantitative Biology* 18:123–131.

Fig. 11. *See 10.*

Fig. 12. No reference.

Fig. 13. V. A. McKusick. 1964. *Human Genetics.* Englewood Cliffs, N.J.: Prentice-Hall.

Fig. 14. C. Pavan and M. E. Breuer. 1952. Polytene chromosomes in different tissues of *Rhynchosciara. Journal of Heredity* 53:149–157.

Fig. 15. M. E. Breuer and C. Pavan. 1955. Behavior of polytene chromosomes of *Rhynchosciara angelae* at different stages of larval development. *Chromosoma* (Berlin) 7:371–386.

Fig. 16. C. B. Bridges. 1935. Salivary chromosome maps. *Journal of Heredity* 26:60–64.

Fig. 17. Th. Dobzhansky and A. H. Sturtevant. 1938. Inversions in the chromosomes of *Drosophila pseudoobscura. Genetics* 23:28–64.

Fig. 18. Th. Dobzhansky and C. Epling. 1944. Contributions to the genetics, taxonomy, and ecology of *Drosophila pseudoobscura* and its relatives. *Carnegie Institution of Washington Publications* 554:1–183.

References and Acknowledgments

Fig. 19. *See* 18.

Fig. 20. C. B. Bridges, E. N. Skoog, and J. C. Li. 1936. Genetical and cytological studies of a deficiency (*Notopleural*) in the second chromosome of *Drosophila melanogaster*. *Genetics* 21:788–795.

B. P. Kaufmann and R .C. Bate. 1938. An X-ray induced intercalary duplication in *Drosophila* involving the union of sister chromatids. *Proceedings of the National Academy of Sciences* 24:368–371.

Figs. 21 and 22. *See* 18.

Fig. 23. Th. Dobzhansky and D. Sokolov. 1939. Structure and variation of the chromosomes in *Drosophila azteca*. *Journal of Heredity* 30:3–19.

Fig. 24. *See* 18.

Fig. 25. No reference.

Fig. 26. Helena Slizynska. 1938. Salivary gland analysis of the white-facet region of *Drosophila melanogaster*. *Genetics* 23:291–299.

C. P. Swanson. 1957. *Cytology and cytogenetics*. Englewood Cliffs, N.J.: Prentice-Hall.

Fig. 27. J. T. Patterson and W. S. Stone. 1952. *Evolution in the Genus Drosophila*. New York: The Macmillan Company.

Fig. 28. Th. Dobzhansky. 1937. *Genetics and the Origin of Species*, 1st ed. New York: Columbia University Press.

Fig. 29. K. Patau. 1935. Chromosomenmorphologie bei *Drosophila melanogaster* und *Drosophila simulans* und ihre genetische Bedeutung. *Naturwissenschaften* 23:537–543.

Th. Dobzhansky and C. C. Tan. 1936. Studies on hybrid sterility. III. A comparison of the gene arrangement in two species, *Drosophila pseudoobscura* and *Drosophila miranda*. *Zeitschrift für induktive Abstammungs- und Vererbungslehre* 72:88–114.

Fig. 30. *See* 29.

H. Burla, A. B. da Cunha, A. R. Cordeiro, Th. Dobzhansky, C. Malogolowkin, and C. Pavan. 1949. The *willistoni* group of sibling species of *Drosophila*. *Evolution* 3:300–314.

Fig. 31. T. R. F. Wright and R. MacIntyre. 1963. A homologous gene-enzyme system, esterase-6, in *Drosophila melanogaster* and *D. simulans*. *Genetics* 48:1717–1726.

Fig. 32. Th. Dobzhansky. 1951. *Genetics and the Origin of Species*, 3d ed. New York: Columbia University Press.

Fig. 33. C. D. Darlington. 1956. *Chromosome Botany*. London: George Allen and Unwin.

Fig. 34. A. G. Avery, Sophie Satina, and J. Rietsema. 1959. *Blackeslee: The Genus Datura.* Vol. 20 of *Chronica Botanica.* New York: Ronald Press.

Fig. 35. *See* 16.

Fig. 36. V. M. Ingram. 1963. *The Hemoglobins in Genetics and Evolution.* New York: Columbia University Press.

C. Baglioni. 1962. The fusion of two peptide chains in hemoglobin Lepore and its interpretation as a genetic deletion. *Proceedings of the National Academy of Sciences* 48:1880–1886.

Fig. 37. *See* 36.

Fig. 38. No reference.

Fig. 39. J. F. Speyer, P. Lengyel, C. Basilio, A. J. Wahba, R. S. Gardner, and S. Ochoa. 1963. Synthetic polynucleotides and the amino acid code. *Cold Spring Harbor Symposium on Quantitative Biology* 28:559–567.

Glossary

allele A given state of a gene. For example, the *white* gene in the vinegar fly has numerous alleles: the normal allele which is responsible for the production of the normal eye pigments, the white allele which fails to produce any pigment, and a host of intermediate alleles (eosin, apricot, blood, cherry, and others) which produce intermediate amounts of pigment.

amino acid Any one of twenty or so molecules that make up proteins. Amino acids are made of carbon, hydrogen, oxygen, nitrogen, and sulfur (in one case).

anemia A blood disorder characterized by insufficient hemoglobin or by an insufficient number of red blood cells.

asexual reproduction Reproduction without recourse to sexual processes. Examples are the ordinary division of bacteria, the budding of yeast, and propagation by tubers in potatoes or runners in strawberries.

base pair A pair of nitrogenous bases (one purine and one pyrimidine) that join the two parallel strands of the DNA molecule.

centromere The region of the chromosome that controls its movement during cell division. Its position on the chromosome also governs the physical appearance (rod-, J-, or V-shape) of the chromosome during cell division.

chromosomal aberration Any abnormal arrangement of the normal chromosome complement caused by chromosomal breakage

163

and reunion. Wild populations of many organisms contain a variety of naturally occurring aberrations.

chromosomal break A break in the chromosome that serves as the basis for the repatterning of the physical structure of chromosomes. Inverted blocks of genes (inversions) and the exchange of chromosome arms (reciprocal translocations) are two commonly observed rearrangements; each requires a minimum of two breaks.

chromosomal exchange The reciprocal translocation of nearly intact chromosome arms. Two breaks, one in each chromosome, are required for this type of exchange.

chromosome One of several bodies composed of protein and nucleic acid (both DNA and RNA) that are found in cell nuclei. Genes, the ultimate determiners of heritable traits, are located on chromosomes.

chromosome, giant An exceptionally large chromosome found in certain tissues (notably the cells of the salivary glands) of various flies (*Diptera*). Giant chromosomes arise through the repeated replication of chromosomal material in the absence of cell division.

chromosome arms Portions of the chromosomes extending from the centromere to the free end. Since centromeres of giant chromosomes tend to disintegrate when cells are prepared for microscopic examination, the most obvious structures in these preparations are chromosome arms.

clone A group of individuals that have arisen by asexual reproduction and, consequently, are genetically identical.

deletion The loss of a small segment from a chromosome. In giant chromosomes a deletion causes the normal, nondeleted chromosome to form a small unpaired loop opposite the deleted segment.

DNA (deoxyribonucleic acid) The chemical that contains genetic information or, more specifically, directions for the synthesis of large and complex protein molecules. DNA is capable of self-replication and, consequently, serves as the vehicle for the transfer of genetic information from parent to offspring.

elements Chromosome arms of different species of *Drosophila* that carry genes whose mutant alleles cause similar abnormalities.

embryology The science of early development, of the transformation of a fertilized egg into an embryonic individual comprising many tissues, organs, and organ systems.

enzyme A protein that catalyzes a certain reaction. Virtually all chemical reactions in living cells are under enzymatic control.

facet One of the hundreds of small surfaces of the compound eye of an insect; each acts as a lens for the simple cylindrical eye beneath it. *Facet* (italicized) is the name of a gene in the vinegar fly whose locus is at band 3C7; mutant alleles of this gene cause irregularities in the development of the compound eye.

gamete A germ cell: sperm (male) or egg (female).

gene The unit of Mendelian inheritance; a segment of DNA responsible, in most instances, for the formation of a particular protein or amino-acid chain.

gene arrangement A term used almost exclusively in relation to giant chromosomes where it refers to the sequence of bands along the length of the chromosome. Chromosomes that differ in gene arrangement can be shown to differ from one another through a series of rearranged segments arising as the result of chromosomal breaks.

gene locus The place on a chromosome at which the gene affecting a certain trait is located. Thus, the locus for the *white* gene in the vinegar fly is band 3C1; the gene actually present at that locus in a particular chromosome may be the wildtype allele (the gene for red eyes) or any one of many mutant alleles including *white,* itself.

genome A term used in reference to all of the genes carried by a single gamete, that is, by a single representative each of all chromosome pairs.

haptoglobin A protein, found in blood serum, whose function seems to be to rid the serum of free hemoglobin.

helix A geometric figure, resembling the spring in a screen-door hinge, formed by the rotation of a line around, and at a constant distance from, a longitudinal axis. (A *spiral,* in contrast, twists about at an ever increasing distance like the mainspring of a clock.)

hemoglobin The protein contained in red blood cells whose function is the transportation of oxygen from the lungs to the various tissues of the body.

homologues The two chromosomes of a pair, or the corresponding genes located on the two chromosomes of a pair. Homologues are related by descent.

hybrid An individual obtained by mating two identifiably different parents; thus, there are interpopulation hybrids, interstrain hybrids, interrace hybrids, interspecific hybrids, and even intergeneric hybrids.

inversion A segment of a chromosome that has had, as the result of two chromosomal breaks and rotation through 180°, its gene order reversed relative to the remainder of the chromosome.

lethal mutation A mutation that can cause the death of its carriers. Dominant lethals kill individuals carrying them in single dose; recessive lethals kill individuals carrying them in double dose.

linkage group A group of genes that are not inherited independently of one another. Each linkage group represents the genes carried by a particular chromosome.

linkage map A sequential order of genes arrived at by counting the offspring of certain controlled matings. The map reveals the extent to which genes of a single linkage group are inherited independently of one another (see **linkage group**).

Malpighian tubule The excretory organ of insects; a tube composed of cells that absorb uric acid from the body fluids and pass it in granular form into the intestine.

microbial genetics The genetics of micro-organisms.

molecular biology A modern branch of biology concerned with the understanding of biological phenomena in molecular terms. Molecular biology arose largely from the fusion of genetics with biochemistry.

mongoloid idiocy An abnormal syndrome in man characterized by physical abnormalities and mental retardation. Its cause in many instances can be traced to the presence in affected individuals of three (rather than two) members of one particular chromosome.

morphology Science dealing with the visible structures of an organism, the developmental history of these structures, and the comparative relations of similar structures in different organisms.

mutant gene (also **mutant allele**) An altered gene that is incapable of carrying out the function of the normal allele.

nucleotide A portion of the DNA molecule consisting of phosphate, a sugar, and a nitrogenous base (a portion that represents a segment of the longitudinal strand and half of the cross-rung of the ladderlike molecule).

phylogeny The evolutionary history of a group of organisms. Inversions and translocations can be used to trace phylogenies.

polymorphism The existence (at reasonable frequencies) of two or more genetically different forms (such as red and silver foxes, or Rh-positive and Rh-negative human beings) in the same interbreeding population.

polypeptide chain A chain of amino acids whose exact synthesis is under the control of DNA. Some proteins consist of only one chain; others are aggregates of several chains.

polyploid An individual having more than two sets of chromosomes.

protein The characteristic substance of life. All proteins are built of smaller molecules called amino acids.

repeat pairing An anomalous pairing of bands in giant chromosomes that reveals the homology of adjacent bands. Of the two possible types—reverse repeats (*abcddcba*) and tandem repeats (*abcdabcd*)—reverse repeats are much more common in *Drosophila*.

reproductive isolation A term referring to the absence of interbreeding between members of different species. There are many mechanisms by which reproductive isolation is achieved; some of these include different courtship patterns, different flowering times, and the choice of isolated habitats.

ribonuclease An enzyme capable of destroying ribonucleic acid (RNA).

RNA (ribonucleic acid) Nucleic acid that has several important roles in the synthesis of proteins. Messenger RNA is a transcript of the message carried by DNA; soluble RNA matches amino acids with certain combinations of bases in the transcript; and ribosomal RNA serves as the physical support for messenger RNA, soluble RNA, and the new protein molecule during the synthesis of the latter.

self-replication The construction by means of directions issued by one molecule of a second molecule exactly like the first. DNA is a self-replicating molecule; it is a double structure, each half of which can direct the formation of its complement.

species A group of individuals reproductively isolated from all other comparable groups. (The taxonomic species used by museum workers is defined somewhat differently: a group of individuals differing so much from other groups that an experienced worker assigns a Latin name to it. This is a species of convenience.)

translocation A chromosomal aberration involving an exchange between different chromosomes (see **chromosomal exchange**).

viral genetics The genetics of viruses and bacteriophages.

x-radiation Radiation produced when high-speed electrons strike a target. X-rays have relatively short wavelengths and, consequently, are high-energy radiations.

Index

[Italicized numbers refer to pages on which illustrations occur.]

Index